Socio Economic Applications in Geographical Information Science

Socio-Economic Applications of Geographic Information Science

Editors David Kidner, Gary Higgs and Sean White

CRC Press
Taylor & Francis Group
Boca Raton London New York

CRC Press is an imprint of the
Taylor & Francis Group, an **informa** business

First published 2002 by Taylor & Francis Group

Published 2020 by CRC Press
Taylor & Francis Group
6000 Broken Sound Parkway NW, Suite 300
Boca Raton, FL 33487-2742

First issued in paperback 2020

ISBN 13: 978-0-367-57856-5 (pbk)
ISBN 13: 978-0-415-27910-9 (hbk)

Visit the Taylor & Francis Web site at
http://www.taylorandfrancis.com

and the CRC Press Web site at
http://www.crcpress.com

Publisher's Note
This book has been prepared from camera-ready copy provided by the editors.

Every effort has been made to ensure that the advice and information in this book is tru
at the time of going to press. However, neither the publisher nor the authors can accept
responsibility or liability for any errors or omissions that may be made. In the case of d
administration, any medical procedure of the use of technical equipment mentioned wi
you are strongly advised to consult the manufacturer's guidelines.

British Library Cataloguing in Publication Data
A catalogue record for this book is available from the British Library

Library of Congress Cataloging in Publication Data
A catalog record for this book has been requested

Contents

Preface

This volume contains papers presented at the 9th Annual GIS Research UK conference held at the University of Glamorgan and co-hosted by the University of Wales, Cardiff. We are proud to declare that this was the most well attended GISRUK conference to date and had the highest number of paper submissions. This stemmed from an increasing interest in GIS Research from both within, and outside, the academic community and reflects a growing maturity in the use of GIS in a number of different sectors. The Local Organising Committee for GISRUK 2001 made a conscious effort to target a broader range of papers to appeal to a wider audience and this led to sessions containing papers arranged according to different policy sectors which together occupied the middle day of the conference. This is reflected in the diversity of papers included in this volume which, to a certain extent, complements those included in last years volume edited by Peter Halls which was largely concerned with the innovative use of GIS in environmental applications. The aim here is to reflect the significant body of innovative research that is being conducted largely in the social sciences. At the same time, we have not neglected important research initiatives in the physical sciences – papers arranged around these themes will be included in special issues edited by the co-chairs of the meeting. Of course, many of the papers presented here transcend such arbitrary, and increasingly fuzzy, boundaries and a key message from the papers presented here and in the other conference outputs, is that many of the techniques developed in either 'domain' are indeed transferable. This, in turn, suggests that the oft-repeated claim that GIS is helping to break down barriers between the physical and social world is being realised through a host of interdisciplinary initiatives.

At this point, it is appropriate to remind readers of the aims of the GISRUK conference, which are:

- to act as a focus for GIS Research in the UK;
- to provide a mechanism for the announcement and publication of GIS Research;
- to act as an interdisciplinary forum for the discussion of research ideas;
- to promote active collaboration amongst UK researchers from diverse parent disciplines; and
- to provide a framework in which postgraduate students can see their work in a national context.

This year's programme, and attendance list, suggests that GISRUK has evolved into something more than just a British forum for GIS research. Approximately one third of the presentations were made up from International contributions, which is very encouraging. We have made a deliberate effort to include contributions from a range of social and physical environments in our conference outputs which, we hope, will add to the international appeal of the volume.

The conference included plenary keynote addresses from Vanessa Lawrence (Ordnance Survey), David Maguire (ESRI), Martien Molenaar (ITC) and Nick Chrisman (University of Washington). We are delighted to include Nick Chrisman's contribution as the opening chapter in this volume, which explores a number of important research challenges addressed by many of the chapter contributors. As in previous years, a prize

(sponsored by the Association for Geographic Information) was awarded tc
paper presented at GISRUK 2001 and this year the prize was given to Anna S
of the University of Newcastle (to our knowledge, the youngest presenter at the
for her research exploring the use of statistical techniques to trace errors when
map lineage. In addition a prize, sponsored by CADCORP Ltd, was given tc
poster presentation and, after much discussion amongst the Steering Comm
hotly contested award was given to Oliver Duke-Williams and John Stillwc
University of Leeds for their poster illustrating their research on developing w
interfaces to access migration and travel to work statistics from the 1981 and
censuses.

As in previous years, GISRUK actively encourages full involvemc
postgraduate and young researchers with the help of substantial registration disco
Bursary awards, both of which are made possible by our sponsors (AGI, RRI
Ordnance Survey). The Young Researchers Forum (YRF) which precedes
conference is a unique component of GISRUK and we would like to tha
members of the national and local organising committee, as well as our keynote
who attended this year's forum. Particular thanks should go to Peter Halls a
Orford who helped put an extremely interesting and varied timetable togethe
year's forum. Feedback from this and previous years' YRFs has been very enc
and it is hoped that they continue as a regular feature of the GISRUK meetings.

The Steering Committee are enormously grateful to the sponsors (listed
xiv) who generously supported GISRUK 2001. We would particularly like
Taylor and Francis and the AGI for sponsoring the keynote presentations this ye
Wileys and Elsevier for their sponsorship of the wine receptions, always a key 1
GISRUK! We would also like to thank the staff at the Glamorgan Business Ce
the Local Organising Committee (listed on page *xiv*), for their support dt
planning of the meeting. The continued support of members of the National O
Committee is much appreciated. A particular thanks should go to Jan Cross and
Bowen for all their help in the day-to-day running of the conference. To all: I
fawr.

Lastly, to keep informed of future GISRUK conferences, please bc
http://www.geo.ed.ac.uk/gisruk/gisruk.html.

Gary Higgs, David Kidner and Sean White
Trefforest
January 2002

Contributors

Kheir Al-Kodmany
University of Illinois at Chicago, UIC, United States of America
E-mail: Kheir@uic.edu

Kate Bowers
URPERRL, Department of Civic Design, University of Liverpool, Liverpool, L69 7ZQ, United Kingdom
E-mail: kjb@liverpool.ac.uk

Spencer Chainey
InfoTech Enterprises Europe, 40-42 Parker Street, London, WC2B 5PQ, United Kingdom
E-mail: schainey@infotech-europe.com

Nicholas Chrisman
Department of Geography, University of Washington, Seattle, WA 98195-3550, United States of America
E-mail: chrisman@u.washington.edu

Jonathan Corcoran
GIS Research Centre, School of Computing, University of Glamorgan, Pontypridd, CF37 1DL, United Kingdom
E-mail: jcorcora@glam.ac.uk

Lynn Dyson-Bruce
Planning Division, Essex County Council, County Hall, Chelmsford, Essex, CM1 1QH, United Kingdom
E-mail: lynn.dyson-bruce@essexcc.gov.uk

Steve Evans
Centre for Advanced Spatial Analysis, University College London, 1-19 Torrington Place, WC1E 6BT, United Kingdom
E-mail: sevans@geog.ucl.ac.uk

Robin Flowerdew
School of Geography and GeoSciences, University of St Andrews, KY16 9AL, United Kingdom
E-mail: r.flowerdew@st-andrews.ac.uk

Brian Francis
Lancaster University, Centre for Applied Statistics, Lancaster, LA1 4YB, United Kingdom
E-mail: b.francis@lancaster.ac.uk

Andrea Frank
Department of City and Regional Planning, University of Wales, Cardiff, Cl
United Kingdom
E-mail: FrankA@cardiff.ac.uk

Martin Frost
School of Geography, Birkbeck College, University of London, 7-15 Gres
London, W1P 2LL, United Kingdom
E-mail: m.frost@bbk.ac.uk

Myles Gould
School of Geography, University of Leeds, Leeds, LS2 9JT, United Kingdom
E-mail: M.Gould@geog.leeds.ac.uk

Andy Hamilton
School of Construction and Property Management, Salford University, Salford,
United Kingdom
E-mail: a.hamilton@salford.ac.uk

Juliet Harman
Lancaster University, Centre for Applied Statistics, Lancaster, LA1 4YB, United
E-mail: j.harman@lancaster.ac.uk

Richard Harris
School of Geography, Birkbeck College, University of London, 7-15 Gres
London, W1P 2LL, United Kingdom
E-mail: r.harris@bbk.ac.uk

Robin Haynes
School of Environmental Sciences and School of Health Policy and Practice, U
of East Anglia, Norwich, NR4 7TJ, United Kingdom
E-mail: R.Haynes@uea.ac.uk

Gary Higgs
GIS Research Centre, School of Computing, University of Glamorgan, Pontypri
1DL, United Kingdom
E-mail: ghiggs@glam.ac.uk

Alex Hirschfield
URPERRL, Department of Civic Design, University of Liverpool, Liverpool,
United Kingdom
E-mail: hirsch@liv.ac.uk

Andrew Hudson-Smith
Centre for Advanced Spatial Analysis, University College London, 1-19 Torrington Place, WC1E 6BT, United Kingdom
E-mail: asmith@geog.ucl.ac.uk

Shane Johnson
URPERRL, Department of Civic Design, University of Liverpool, Liverpool, L69 3BX, United Kingdom
E-mail: s.d.johnson@liverpool.ac.uk

Mandy Kelly
Lancaster University, North-West Regional Research Laboratory, Department of Geography, Lancaster, LA1 4YB, United Kingdom
E-mail: Mandy.Kelly@lancaster.ac.uk

David Kidner
GIS Research Centre, School of Computing, University of Glamorgan, Pontypridd, CF37 1DL, United Kingdom
E-mail: dbkidner@glam.ac.uk

Peter Lee
Centre for Urban and Regional Studies, University of Birmingham, JG Smith Building, Pritchatts Road, Edgbaston, Birmingham, B15 2TT, United Kingdom
E-mail: p.w.lee@bham.ac.uk

Andrew Lovett
School of Environmental Sciences, University of East Anglia, Norwich, NR4 7TJ, United Kingdom
E-mail: A.Lovett@uea.ac.uk

Brendan Nevin
Centre for Urban and Regional Studies, University of Birmingham, JG Smith Building, Pritchatts Road, Edgbaston, Birmingham, B15 2TT, United Kingdom
E-mail: b.nevin@bham.ac.uk

Scott Orford
Department of City and Regional Planning, University of Wales, Cardiff, CF10 3WA, United Kingdom
E-mail: OrfordS@cardiff.ac.uk

Svein Reid
Department of Geography, University of Edinburgh, Drummond Street, Edinburgh, EH8 9XP, United Kingdom

Andrew Schuman
Local Government Chronicle Elections Centre, University of Plymouth, Plymc
8AA, United Kingdom
E-mail: Andrew.Schumann@bris.ac.uk

Carsten Schürmann
Institute of Spatial Planning, University of Dortmund, August-Schmidt-Str-6,
Dortmund, Germany
E-mail: cs@irpud.rp.uni-dortmund-de

Darren Smith
Department of Geography, School of the Built Environment, University of .
Brighton, BN2 4GJ, United Kingdom
E-mail: D.Smith@brighton.ac.uk

Neil Stuart
Department of Geography, University of Edinburgh, Drummond Street, Edinbu
9XP, United Kingdom
E-mail: ns@geo.ed.ac.uk

Gisela Sünnenberg
School of Environmental Sciences and School of Health Policy and Practice, l
of East Anglia, Norwich, NR4 7TJ, United Kingdom
E-mail: G.Sunnenberg@uea.ac.uk

Ahmed Talaat
Institute of Spatial Planning, University of Dortmund, August-Schmidt-Str-6, ₄
Dortmund, Germany
E-mail: at@irpud.rp.uni-dortmund-de

Helena Titheridge
The Bartlett School of Planning, UCL, 22 Gordon Street, London, WC1H 0QI
Kingdom
E-mail: h.titheridge@ucl.ac.uk

Nigel Trodd
Telford Institute of Environmental Systems (TIES), School of Environment
Science, Salford University, M5 4WT, United Kingdom
E-mail: n.m.trodd@salford.ac.uk

Andrew Ware
School of Computing, University of Glamorgan, Pontypridd, CF37 1DL
Kingdom
E-mail: jaware@glam.ac.uk

Sean White
Department of City and Regional Planning, University of Wales, Cardiff, Cardiff CF10 3WA, United Kingdom
E-mail: WhiteSD@cardiff.ac.uk

Fulong Wu
Department of Geography, University of Southampton, Southampton, S017 1BJ, United Kingdom
E-mail: F.Wu@soton.ac.uk

Chris Young
URPERRL, Department of Civic Design, University of Liverpool, Liverpool, L69 3BX, United Kingdom
E-mail: youngca@liv.ac.uk

Xiaonan Zhang
Telford Institute of Environmental Systems (TIES), School of Environment and Life Science, Salford University, M5 4WT, United Kingdom
E-mail: X.Zhang1@pgr.salford.ac.uk

GISRUK Committees and Sponsors

GISRUK National Steering Committee (as of GISRUK 2

Steve Carver	University of Leeds	steve@geog.leeds.ac.uk
Jane Drummond	University of Glasgow	jdrummond@geog.gla.a
Dave Fairbairn	University of Newcastle	dave.fairbairn@newcas
Bruce Gittings (Chair)	University of Edinburgh	bruce@geo.ed.ac.uk
Peter Halls	University of York	pjh1@york.ac.uk
Gary Higgs	University of Glamorgan	ghiggs@glam.ac.uk
Zarine Kemp	University of Kent	zk@ukc.ac.uk
David Kidner	University of Glamorgan	dbkidner@glam.ac.uk
Andrew Lovett	University of East Anglia	a.lovett@uea.ac.uk
David Miller	The Macaulay Institute	mi008@mluri.sari.ac.uk
George Taylor	University of Glamorgan	getaylor@glam.ac.uk
Steve Wise	University of Sheffield	s.wise@sheffield.ac.uk
Jo Wood	City University	jwo@soi.city.ac.uk

GISRUK 2001 Local Organising Committee

David Kidner (Co-Chair)	Linus Mofor	Nathan Thomas
Gary Higgs (Co-Chair)	Andrew Morris	Andrew Ware
Jonathan Corcoran	Scott Orford	Mark Ware
Chris Jones	Alun Rogers	Chris Webster
Marc Le Blanc	George Taylor	Sean White
Mike Lonergan	Malcolm Thomas	

GISRUK 2001 Sponsors

The GISRUK Steering Committee is enormously grateful to the following orga who generously supported GISRUK 2001:

The Association for Geographic Information (AGI)	Blackwell Publis
Cadcorp	Edina
Elsevier Science	ESRI (UK) Ltd
Geo-Europe	Ordnance Survey
Oxford University Press	Pearson Educatie
The Quantitative Methods Research Group of the RGS/IBG	RRL.Net
Taylor & Francis Ltd	John Wiley and

1

Introduction

Gary Higgs, David B. Kidner and Sean D. White

1.1 SOCIO-ECONOMIC APPLICATIONS IN GEOGRAPHICAL INFORMATION SCIENCE

In a previous volume in this series papers were included that demonstrated the use of innovative research techniques in environmental studies (Halls, 2000). The aim of this volume is to include papers presented at GISRUK 2001 that illustrate the state-of-the-art in researching socio-economic applications with GIS. In so-doing, we recognise that the intertwined relationships between the physical and human world mean that such arbitrary divisions are not straightforward. There are clearly environmental aspects to the majority of papers that follow, but at the same time, given the quality of papers presented at the conference, we have taken this opportunity to demonstrate the dynamic nature of GIS research in what would traditionally be called 'human geography'. A number of texts have been concerned with focusing on the use of GIS in socio-economic applications (*e.g.* Martin, 1996) and a number of others contain extensive overviews of the use of GIS in different policy sectors (*e.g.* Longley *et al.*, 1999). There have also been texts that have examined the use of GIS in application areas such as crime (Goldsmith *et al.*, 2000; Hirschfield and Bowers, 2001), health (Gatrell and Loytonen, 1998; Hay *et al.*, 2000) and transport (Thill, 2000). Previous volumes in the *Innovations in GIS* series have variously contained papers related to such topics and readers are advised to consult these to get a flavour of the significant research developments that have taken place since the first GIS Research UK conference held in Keele in 1993. However, to date, no one volume has been given over to solely highlighting the use of up-to-date GIS-based techniques in a range of socio-economic applications. In this volume we redress this 'gap'.

In the following chapter, **Nicholas Chrisman** examines how social issues are embedded in the different stages of GIS development. In particular, he examines the limitations of technological determinism from a GIS perspective and draws on wider, more theoretical literatures, to call for a greater understanding of the interactions between people, organisations and such technologies. This is illustrated with a number of examples that draw attention to the role of software vendors in influencing GIS users in their work from a number of different, but inter-connected, perspectives. This, in turn, points to the potential benefits to be gained from greater participation of users in software development processes and their continued independence and objectivity in GIS research. This provides a valuable introduction for many of the chapters in this volume which are largely concerned with enhancing existing software tools in innovative ways to address social applications.

The rest of the book is composed of seventeen chapters, divided into five parts, which focus on the use of GIS in different sectors. In the first section, we focus on the integration of analytical techniques with GIS in order to investigate spatial patterns of

crime incidence. GIS is increasingly being recognised as a vital tool in the expl
spatial trends, in the prediction of crime events and in the evaluation of crime p
measures. In the UK, much of the impetus for the use of GIS in crime an
resulted from legislation such as the Crime and Disorder Act (1998) which ma
crime and disorder audits a statutory requirement. As a result, GIS has t
extensively in the preparation of maps and in preliminary analyses of spatial dat
in the majority of police authorities. In addition, the Home Office document "
on Statutory Crime and Disorder Partnerships" (Stationary Office, 1998) has er
the potential for GIS in extracting patterns from audits. One of the recommen
the guidance documents was that agencies such as local authorities, police a
services need to work together to produce such audits. Together with the
community safety strategies for local areas, GIS is seen as an ideal tool to pe
'joined-up' working. However, many of these authorities are in early stages in
development and the 3 chapters in this section, illustrate how GIS can be used
advanced way to address a number of current strands in researching crime patter

The second section of the book contains 4 chapters that focus on the use
planning tasks. Planning departments in the UK tend to be the lead departme
majority of local authority GIS implementations having traditionally been majo
spatial information and paper maps (Campbell and Masser, 1995). Howeve
having a relatively long history of use within the profession, GIS continues to t
predominantly to address relatively low level, operational and routine tasks, cont
their use in more strategic contexts which has traditionally been very limited (C
1999). The reasons for this are complex and varied, but the contributions contain
section, point the way towards the more innovative use of GIS in areas such as u
based tools in public participation and collaborative visual assessment. One sec
GIS has made less in-roads is in the study of housing issues. Once again, GIS
used as an operational tool in a significant number of agencies in their
management functions but less research has been conducted, certainly in the UK
in using GIS to examine changing demand-supply relationships through the integ
spatial analytical techniques. Clearly, there are many other application areas wh
potential concern to planners, and arguably many of the contributions in the othe
of the book could have been included here, but these chapters begin to examin
contribution GIS can make in more strategic contexts.

In the third section, contributions focus on urban dimensions and the three
contained herein variously describe the use of GIS in urban regeneration, in e
spatial trends in socio-economic patterns and in exploring intra-urban variations
prices. GIS has been used extensively in the latter to examine the impacts, for ex
access to urban services and facilities on house prices (*e.g.* Orford, 1999). A
visualisation aspects of GIS technologies come to the fore here. One of th
research strands pre-occupying many research groups relates to the use o
measuring access to facilities in urban and rural contexts and in section four, w
three chapters which although focusing on such issues from a rural perspective,
methodologies that could easily be adapted for urban scenarios. In particul
chapters draw attention to the limitations of simplistic measures based on '
distances by demonstrating the potential to incorporate travel times into such i
Further research is on-going to include information from public transport timeta
such analyses and this continues to be a fruitful area for further research. The i
importance of Global Positioning Systems (GPS) in such studies is also likely
vital component of future papers given at the annual conference.

In the last section of the volume, we focus on studies of the use of GIS i
aspects of socio-economic research. GIS has been used extensively to develop i

deprivation in both urban and rural contexts (see Higgs and White, 2000 for examples of the latter). However, here we are concerned with the derivation of measures using non-census based sources, or more accurately lifestyle databases, that tend not to be made available at detailed spatial scales. Such datasets have real potential for researchers concerned with investigating socio-economic patterns in inter-censal years as well as those concerned with a wider range of urban issues (Longley and Harris, 1999). Other chapters in this section are concerned with using GIS to investigate aspects of electoral geography and in the development of indices of peripherality. The final chapter provides a detailed account of how we can begin to investigate current uses of GIS in researching health issues within the UK National Health Service using multi-method approaches. Although we acknowledge there are many other application areas that are not covered in this volume, and that many of the chapters could easily have been placed in a number of sections, we believe that these contributions are both illustrative of the types of research initiatives underway in the UK and beyond, and that many of the techniques described are transferable to other application areas which have not been included this time round.

1.2 GIS AND CRIME

In Chapter 3, **Spencer Chainey**, **Svein Reid** and **Neil Stuart** highlight techniques for describing clusters of crime patterns or so-called 'hotspot maps'. They contrast a number of approaches to creating continuous surface maps according to different criteria including ease of use and interpretation, relevance for particular crime data sets and visual appearance. Often despite the use of sophisticated techniques to create such surfaces, it is ultimately up to the user to decide when a cluster can be identified as a hotspot. To address such issues, the authors apply point pattern analysis, and in particular local tests of spatial association as well as spatial autocorrelation, to create surfaces for, in this case, four different categories of crime over a 3 month time period. They contrast different techniques for defining thresholds when applied to these surfaces and compare the outcomes for different types of crime. As Chainey *et al.* suggest, the use of space-time methods to explore temporal clusters could prove a useful addition to the range of techniques available to agencies and there are a number of research initiatives underway to apply such techniques (*e.g.* Ratcliffe, 1998).

Chris Young, **Alex Hirschfield**, **Kate Bowers** and **Shane Johnson** explore another application area for GIS in relation to crime analysis in Chapter 4, namely that of evaluating the success, or otherwise, of crime prevention measures using the case study of an area of Liverpool, Merseyside. Although, based on a relatively short run of data post-intervention, they examine the impacts of alleygating (installing gates at the entrances of small streets at the rear of premises) as part of a series of burglary reduction projects. GIS has significant potential here to investigate the type, nature, and extent of crimes in relation to interventions such as CCTV and Neighbourhood Watch schemes but can also present some interesting challenges for researchers faced with using data sets at a range of temporal and spatial scales. This is further complicated by the use of incompatible spatial units such as police beat areas which do not neatly correspond to other administrative geographies. Further concerns relate to data confidentiality and in particular, the use of spatially disaggregate data on victim incidences.

New methods of analysing crime data are addressed in Chapter 5 by **Jonathan Corcoran** and **Andrew Ware** who describe a project currently underway with a police force in Wales which is concerned with exploring the use of Artificial Neural Networks (ANN) in examining crime incidence. Their paper provides a preliminary analysis of the use of such techniques in predicting crime patterns in Cardiff, South Wales. A significant

amount of work is needed to pre-prepare the data for such an analysis and the
the issues that arise from the use of such detailed data sources before outlining
of the techniques used in these contexts. Finally, the authors re-iterate the polic
of such an approach to crime analysis.

1.3   GIS AND PLANNING

In the second section of the book, we are concerned with showing the potential
a range of planning tasks. There is increasing interest in using GIS within grou
making processes to address community concerns and preferences, and in C
Kheir Al-Kodmany describes the potential for web-based GIS tools in such
participation exercise. Previous research has demonstrated the use of such soft
in, for example, 'Planning for Real' collaborative exercises in UK contexts (*e.g.*
et al., 2000). Such studies have highlighted the importance of designing a
interfaces in community-based GIS studies to take into account the findings from
into cognitive aspects of using screen-based images and maps. This study
concerned with experimenting with prototype interfaces as part of a wider study
the use of web-based tools for community based initiatives for lo
neighbourhoods in Chicago. Residents are asked to comment on the visual appe
their communities through various types of interface. Further research is r
compare the different approaches to interface design in a user environment but
has highlighted the types of advances that have been made in designing web-b
in public participation exercises.

The use of Internet based GIS tools in collaborative planning is also dem
in Chapter 7 by **Andrew Hudson-Smith** and **Steve Evans** who are concer
developing 3D models of urban form in London. Research at the Centre for
Spatial Analysis (CASA) has used basic building blocks and other locational f
2D maps to create photo-realistic 3D block models in networked environments
areas of the city. They critically review some of the data sets needed to cr
models including comparatively new data sets such as LIDAR (Light Dete
Ranging) images. They also highlight some of the drawbacks in creating such 3
within existing GIS software before advocating an approach based on lin
enabled GIS with 3D modelling systems. Users of such systems can query attr
objects within such models, hyperlink to other web-sites or conduct fly- or walk-
and the authors conclude by demonstrating a prototype in one London borougl
research is planned to refine this model and to test its usefulness in practical
tasks such as in regeneration initiatives but the approach adopted has significant
in addressing community concerns.

The visual aspects of GIS are also highlighted in Chapter 8 by **Lynn Dys**
who describes the use of Historic Landscape Assessment (HLA) to examine
diversity in a study area in the East of England. GIS has shown great potential
planning process where development may impinge on land which may have s
historical and cultural significance. This study has also drawn attention to the
consistent meta-data relating to data holdings and the not insignificant amoun
that is involved in collating such data from a range of agencies. Finally Dys
provides examples of how this approach has been applied in the study regio
iterates the advantages of GIS vis-à-vis traditional approaches to handling such d

The last paper in this section by **Peter Lee** and **Brendan Nevin** (Cha
concerned with the use of GIS in assessing the demand for housing in local autl
the UK and, in particular, with the potential for such approaches in targeting r
To date, much of the use of GIS in housing departments has been to man

authority or social housing stock. This research points the way to a more proactive use of GIS in helping to combat social exclusion in a number of urban authority areas. A key element has been the derivation of suitable indices of poverty in such contexts and the use of administrative data in identifying areas at risk of low demand. This, in turn, has been tested against data relating to housing turnover, transfer requests and housing stock condition to examine the appropriateness of such indices.

1.4 GIS AND URBAN APPLICATIONS

In the third section of the book, we are primarily concerned with studies that have specifically involved the use of innovative GIS-based techniques in urban contexts. Clearly, there is overlap here with some of the issues arising from the papers in Section 2, but the aim has been to include papers which have a broader focus. In Chapter 10, **Xiaonan Zhang**, **Nigel Trodd** and **Andy Hamilton** outline a theoretical framework relating to the application of Geographical Visual Information Systems (GVIS) in urban regeneration. They describe the use of such frameworks in relation to stages in the public participation process and document the advantages of such an integrated approach in relation to a pilot area in Salford in the north of England.

In Chapter 11, **Andrea Frank** describes how spatial autocorrelation can be used to identify distinctive areas within US cities. Using census data to identify clusters of areas based on socio-demographic characteristics, the aim has been to examine the hypothesis that local political forces influence residential patterns through land use, taxation and planning policies and that these can be identified through the use of spatial analytical approaches. This has involved a loose coupling approach in order to integrate autocorrelation techniques with a commercially available GIS and applied to cities with contrasting political structures and population characteristics. This, in turn, has drawn attention to key methodological concerns facing researchers in this field, for example, in relation to the definitions of *neighbourhoods*. Frank reiterates the policy implications of such research and the applicability of such techniques in other environments.

The subject of the importance of detailed, spatially disaggregate data sources in studies of socio-economic variations, is returned to by **Fulong Wu** in Chapter 12. Using the case study of Shanghai, contrasts are presented to the situation in the developing world where such data is generally not available. A high resolution spatial data set is created with which to study intra-urban spatial variations in property prices. Recent approaches have investigated the incorporation of locational attributes into hedonic regression models to analyse the determinants of local house prices. In this study, residuals from regression models (having controlled for structural variables of properties) are interpolated from sample points in order to examine the spatial distribution across the city and to gauge the importance of environmental factors on house prices. Such techniques offer significant benefits in the data-poor contexts of some developing countries.

1.5 GIS AND RURAL APPLICATIONS

The use of GIS research in rural contexts is explored in the fourth section, which includes three chapters on analysing various aspects of rural service provision. In Chapter 13, **Andrew Lovett**, **Gisela Sünnenberg** and **Robin Haynes**, describe the use of GIS-based techniques to examine the accessibility of health services, in this case General Practitioner (GP) surgeries in Norfolk, in the east of England. Traditionally such studies have used straight-line distances to investigate such access issues. This is fundamentally flawed both

because of the state of the road network and the reliance of some sectio
community on public transport, which in recent decades has been in decline in s
areas. In this study, the researchers are concerned with taking into account the a
of such services when examining access for those on patient registers. In parti
demonstrate the potential for GIS-based techniques when examining the impli
recent investments in public transport provision within communities. Their s
points the way to identifying communities that are still ill-served by existing
and those that should be targeted for more investment in public or community
schemes.

In Chapter 14, **Mandy Kelly, Robin Flowerdew, Brian Francis a**
Harman continue the theme of using GIS to measure access for rural populatie
prime concern in this paper, however, is to demonstrate how GIS can be used in
allocation measures such as the Standard Spending Assessments (SSAs). There i
that rural areas are losing out under existing funding mechanisms because of the
take into account the unique set of circumstances facing service providers in sv
This research proposes a methodology whereby the remoteness of such areas ca
into account through the use of GIS-based measures of travel times which con
nature of the road network and the time taken to access such areas. The researc
attention to the financial implications of incorporating such measures into
sparsity measures for authorities in England in terms of the 'winners' and '
resource allocation. The results are not as clear-cut as first thought which pre
researchers to call for more research into deriving measures which reflect the
providing services for dispersed populations in remote areas.

The theme of examining the potential for GIS in the area of service pre
rural areas is also addressed in Chapter 15 by **Helena Titheridge**. This researc
by the EPSRC Sustainable Cities programme, is concerned with examining the
for using GIS-based models to reduce travel to services and hence energy con
The ESTEEM (Estimation of Travel, Energy, and Emissions Model) has been a
Gloucestershire in the west of England in order to compare travel patterns under
of policy scenarios relating to service and housing provision. This is run as an
to an existing commercially available GIS package. This is based on a gravi
which is used to examine travel patterns based on variables such as car owner
existing transport networks and the impacts of new developments on fuel con
and emissions are modelled. In this way, the policies for a typical local authori
monitored both for the authority as a whole or for individual settlements. This
has drawn attention to the limitations of existing data sources used to calibr
models. Many authorities do not routinely collect data on changes in the qua
quality) of facilities and a significant amount of effort is needed to collate data
modelling efforts. This study has drawn attention to the limitations of those data
do exist before modelling the transport implications of different policy
involving the location of housing developments and facilities. This, in turn, h
attention to the role of public transport in rural areas.

1.6   GIS IN SOCIO-ECONOMIC POLICY

The final section of the book contains four chapters that are concern
demonstrating the potential for GIS in addressing a variety of socio-economic
Chapter 16, **Richard Harris** and **Martin Frost** highlight the limitations of usin;
(largely) census based measures at relatively coarse scales for measuring in
variations in socio-demographic characteristics. At the same time, it is recog
existing area based measures need to be refined to take on board the fact t

deprived households are located in otherwise affluent areas and vice versa. Many private sector organisations are developing rich data sets on individuals/households based on consumer patterns that have great potential for describing such patterns at finer, and more flexible, spatial scales (Longley and Clarke, 1995). The use of one such spatially disaggregated data set, for the London borough of Brent, is described in this contribution in order to map the social fragmentation that exists within a 'typical' local authority area. The authors conclude by drawing attention to the potential use of such data sets in relation to recent research concerned with deriving indices of deprivation based on a wider range of measures across the UK and in relation to Government initiatives based around the concept of neighbourhoods.

There has been concern expressed regarding the apathy of voters in recent European and General elections in the UK. In Chapter 17, **Scott Orford** and **Andrew Schuman** investigate the use of a combined GIS-spatial analytical approach to examine the factors determining voter turn-out using the example of voting patterns at a local Council election in a Bristol ward. Specifically they examine the relative influences of geographical factors such as access to polling booths as well as household characteristics using electoral registers showing those individuals who had voted. These were geo-coded via the postcode and a lifestyles database used to examine intra-ward socio-economic patterns. It is then possible to examine the importance of distance to polling stations in relation to such patterns and make some preliminary attempts to suggest ways in which turn-out could potentially be improved, *e.g.* re-locating stations or by altering their catchment areas.

In Chapter 18, **Carsten Schürmann** and **Ahmed Talaat** broaden the discussion to an all-Europe level by describing a project which has been concerned with developing a European Peripherality Index using GIS-based methodologies. As well as distance factors such indices need to take account of the economic potential of the regions and the paper includes some preliminary examples of the types of measures that can be developed. This, in turn, has brought to the attention of policy makers the need for standardised data sets across the EU with which to examine spatial patterns. In order to assess the peripherality of regions travel times by different modes of transport are used in the calculations and combined with economic measures of the various regions (such as Gross Domestic Product) and applied to the NUTS Level 3 spatial units. These measures are compared visually and the policy implications of using various indicators of peripherality discussed. Finally the authors suggest some refinements to the methodology, such as the incorporation of measures based on other transport modes (*e.g.* rail, air), in order to support the assessment of EU policies with respect to peripherality and cohesion.

The final chapter in the volume by **Darren Smith**, **Gary Higgs** and **Myles Gould** describes an on-going project which is concerned with examining the factors influencing the take-up of GIS by health organisations in the UK. This involves the use of mixed method methodologies and the researchers discuss the rationale for such an approach, the expected outputs from the study and the wider implications of the research.

1.7   CONCLUSIONS

In this introductory chapter, we have attempted to outline the main themes to be drawn from the contributions included in this volume. Although we have focused here on the use of innovative GIS-based research techniques in a variety of socio-economic applications, many of the techniques highlighted have relevance for those concerned with addressing wider environmental concerns. We contend that many of the issues discussed in this summary chapter clearly have resonance for those working in other sectors that we have not been able to include in the book. In addition, although we have tended to technical issues of implementation many overarching organisational issue, common across all sectors. Convincing senior managers and politicians of the contributions GIS can make in addressing crucial social and environmental i time of restricted research budgets is of prime concern to many researchers. hoped, that by demonstrating the policy benefits of using GIS in conjunction spatial analytical approaches, this volume has made a contribution to such lauda

1.8 REFERENCES

Campbell, H. and Masser, I., 1995, *GIS and Organizations: How Effective a Practice?* (London: Taylor and Francis).

Gatrell, A. and Loytonen, M. (editors), 1998, *GIS and Health.* (London: T Francis).

Gill, S., Higgs, G. and Nevitt, P., 1999, GIS in planning departments: prelimina from a survey of local authorities in Wales. *Planning Policy and Practice,* 341-361.

Goldsmith, V., McGuire, P.G., Mollenkopf, J.H., Ross, T.A. (editors), 2000, *Crime Patterns: Frontiers of Practice.* (London: Sage).

Halls, P.J., (editor), 2000, *Spatial Information and the Environment: Innovatio 8.* (London: Taylor and Francis).

Hay, S.I., Randolph, S.E., and Rogers, D.J. (editors), 2000, *Advances in Par Remote Sensing and Geographical Information Systems in Epidemiology.* (S Academic Press).

Higgs, G. and White, S.D., 2000, Alternative indicators of social disadvantag communities: the example of rural Wales. *Progress in Planning,* **53**(1), pp. 1

Hirschfield, A. and Bowers, K. (editors), 2001, *Mapping and Analysing Cri Lessons from Research and Practice.* (London: Taylor and Francis).

Kingston, R., Carver, S., Evans, A. and Turton, I., 2000, Web-based public par geographical information systems: an aid to local environmental decisio *Computers, Environment and Urban Systems,* **24**(2), pp. 109-125.

Longley, P. and Clarke, G. (editors), 1995, *GIS for Business and Service* (London: GeoInformation International).

Longley, P.A., Goodchild, M.F., Maguire, D.J. and Rhind, D.W. (editor *Geographical Information Systems: Principles, Techniques, Management I Applications (2nd edition).* (Chichester: Wiley).

Longley, P.A. and Harris, R.J., 1999, Towards a new digital data infrastructure analysis and modelling. *Environment and Planning B,* **26**, pp. 855-878.

Martin, D., 1996, *Geographic Information Systems: Socio-Economic App* (London: Routledge).

Orford, S., 1999, *Valuing the Built Environment: GIS and House Price* (Aldershot: Ashgate).

Ratcliffe, J.H., 1998, Aoristic crime analysis. *International Journal of Geo Information Science,* **12**(7), pp. 751-764.

Stationary Office, 1998, *Guidance on Statutory Crime and Disorder Par* (London: Home Office Communication Directorate).

Thill, J-C. (editor), 2000, *Geographic Information Systems in Transportation* (New York: Permagon).

2

Revisiting fundamental principles of GIS

Nicholas Chrisman

2.1 INTRODUCTION

In 1987, I presented *'Fundamental Principles of GIS'* at Auto-Carto 8 (Chrisman, 1987a).
I want to review the content of this paper, and update it with more recent research. First,
the paper demonstrates that the GIS community has always been embedded in social
issues. We have had our own little version of the Science Wars, and like most wars, both
sides have lost more than they have gained. There are good reasons to ask probing
questions about the human values of any technical system, and I hope we can ask probing
questions about GIS.

On balance, the views in the 1987 article might be asking good questions, but they
seem a bit formulaic in the kinds of remedies presented. Research on 'Society and GIS'
has progressed. I will present some of the exciting new directions, some of them
developed from a closer reading of the literature on the social studies of technology and
science (STS). The most exciting one for me are the ones that are directly linked to the
most seemingly technical details. I want to develop in particular the way that the current
software industry seems intent on 'configuring the user' – rather than developing new
modes of interaction.

2.2 RETURNING TO 1987

In 1987, I presented a paper at Auto-Carto 8 under the title *Fundamental Principles of
GIS* (later published in *Photogrammetric Engineering and Remote Sensing* under the
more descriptive but less punchy title: *Design of GIS based on social and cultural goals*
(Chrisman, 1987b). I do not intend to indulge in some form of nostalgia, but rather to use
the paper I wrote fourteen years ago as a lens to examine the current state of geographic
information systems research.

First, let us return to that time period to set some of the scene. As before and after,
the GIS community was marvelling at rapid expansion. In the autumn of 1986, the British
hosted an unexpectedly overflowing Auto-Carto London (the surplus revenues from that
event may have played a role in many graduate student bursaries ever since, and perhaps
also played a role in founding GISRUK). The first GIS textbook (Burrough, 1986)
appeared at that conference. As Director of the next Auto-Carto, I adjusted my deadline
for abstracts to attract good papers from authors at the London event.

The state of the art in GIS in 1986/87 was (retrospectively) at a high po
acceptance of topological principles to organize software. For example, Interg
developed TIGRIS (Herring, 1987), an object-oriented system using a
topological data structure indexed using R-trees. Since this had been my origin
at earlier Auto-Cartos, you might expect me to have been pleased. Actually, I ha
turned towards a more decentralized model that I articulated more careful
Fundamental Principles paper.

As I sat down to write my own paper for Auto-Carto 8, I knew I did no
write a narrow technical contribution. It was an opportunity to make
comprehensive statement. It is important to remember how important these co
were to the community at that time. The *International Journal of GIS* had just a
that it would begin in 1987, but we mostly relied on conference proceedin
medium to express ideas publicly. While it may seem academically retrograde
organized meeting can often accomplish much more than the most rigorous peer-
journal can ever pull together.

The basic point of the paper (and the sentence most cited since) stated
technology "must be accountable economically, but also politically, socially,
ethically" (Chrisman, 1987b, p. 1367). Such a noble sentiment might seem s
non-controversial in the current epoch in which all undertakings are subject t
questioning, but at the time this was somewhat novel, and even slightly shockin
committed to the technical agenda. The paper proceeded to spend most of its e
the design of databases. It started with an allusion to the earlier era of raster/vect
as a dead issue, without much of an apology for the role I played in it. It tried t
the reliance on user-need studies because they tended to ratify the current stat
then tried to implicate the classical communications model as a source of the
and to present a culturally and historically embedded alternative. Probably not
figured out the diagram that I offered to explain how institutions manage t
holdings are the historical result of the interactions of people and their envi
When I reviewed this diagram in 1992, I found it overly structuralist (Chrism
Any situation involving long-term organizations and the people acting within t
have to deal with the tension between structure and agency, though I did not artic
connection to the social science literature on that subject originally. The artic
what would now appear to be a quaint foray into the lack of objectivity of ge
information. The framing of these ideas still needs to be worked out, it has onl
the past few years that I have begun to locate some ways to address these probl
below).

If the paper had a positive suggestion, it was in the form of a fairly
approach to requirements analysis through the analysis of *mandates* ass
custodians. To some extent, the GIS coordination efforts of the past fourteen y
confirmed this approach. It is not too surprising that Wisconsin adopted this
nearly verbatim, but it also influenced the US Federal Geographic Data Commit
number of other similar efforts. In most cases, the process was somewhat
Instead of rethinking the content of databases from the start using the mand
guide, it seems that the existing agencies declared themselves custodians thr
political process of turf battles, then designed databases that ratified the di
labour. In either case, the result is a more decentralized and collaborative fram
cooperation than was being advanced in that period (National Research Council
example). My voice was just one of many, and the idea of cooperatio
contributors had many origins and many supporters. In any case, it took years to
movements take root in the community. During these years many other forces i
so it is pointless to argue which elements started anything.

The final point of the paper was to argue that *equity* was a more important goal than the measures of efficiency that had dominated the technical arguments. "Geographic information systems should be developed on the primary principle that they will ensure fairer treatment of those affected by the use of the information" (Chrisman, 1987b, p. 1370). I turned this argument into a slight dig at the raster world by arguing that the system should retain the units of interest, not impose some external set of arbitrary units. I ended on a reflexive note, arguing that my earlier work (Chrisman, 1975) proposing an integrated topological database "was flawed because it centralizes definitions" (Chrisman, 1987b, p. 1370). The argument was based on displacing authority to a technical elite that should be more accessible to the political and administrative process. The final sentences demonstrate the ringing polemic of the paper:

"... the search for technical efficiency must not be allowed to overturn political choices without careful examination through the political process. The true challenge is to use the increased sophistication of our automated systems to promote equity and other social ends which will never fit into a benefit/cost reckoning. I am convinced that the future of geographic information systems will lie in placing our technical concerns in their proper place, as serious issues worthy of careful attention. These technical concerns must remain secondary to the social goals that they serve." (Chrisman, 1987b, p. 1370.)

2.3 SO WHAT?

Perhaps it is somewhat soothing to know that conferences fifteen years ago, back in the dark ages of computing, could include such stirring sentiment. No doubt conference speakers have proclaimed equally lofty goals before and since. If my goal were simply to provide a bit of moral uplift after a few days of technical detail, I could simply repeat the main points I delivered in 1987. But, fair listeners, you will not be so easily rid of me. As I took apart my own work from 1974 and 1975 in the 1987 talk, I will now consider how my research direction has changed, and how the prospects for the future seem somewhat different. I will embark on this project with a lot more assistance than I had in the prior enterprise, in part due to an expanded group of GIS researchers who study 'Society and GIS' in various forms, and in part because I have spent the last seven years reading heavily in the interdisciplinary field of Science and Technology Studies (STS). This work leads me to quite a different set of prescriptions from those advanced at Auto-Carto 8. In particular, I want to demonstrate one theme begun in the earlier paper: there are traces of the social in the deepest and most technical parts of a GIS. But first, I need to deal with the discovery of GIS by the rest of the geography research community and the Science Wars that spilled into our isolated world.

2.4 SCIENCE WARS: SHIPS PASSING IN THE NIGHT IN K-D SPACE

Part of the value of returning to 1987 is that it is prior to the wave of criticism that followed. It is not that my paper was the first to connect technical details to social concerns (*e.g.* see Bie, 1984), but that the nature of discourse changed radically.

At first, there was a kind of nervous criticism, most clearly articulated in the intemperate newsletter column of the President of the AAG (Jordan 1988). Professor Jordan, an historical geographer from the 'exceptionalist school', worried that GIS might 'swamp' the discipline and displace focus from the 'theoretical core'. Brian Harley (1989) brought in a post-modernist critique, mostly of maps, but with some mention of the GIS movement. GIS became the sticking point for geographers who wanted to

complain about relationships of power and representation. Perhaps the mc
rhetoric (Taylor & Overton, 1991; Smith, 1992) was enflamed by the 'best de
good offense' strategy of Stan Openshaw (1991). When Jerome Dobson had pub
original 'automated geography' (Dobson, 1983), the commentary had bee
exclusively from cartographers and GIS insiders. When he renewed his vision
years (Dobson, 1993), the commentary included a lot sharper criticism (Pickl
Viewed from inside, it might seem that the GIS movement was being attacked
successful, but the criticism was not simply a reaction to GIS, it is a small part
larger intellectual movement that swept across the humanities and social science
GIS folk felt attacked, the science community reacted, in even more flamboy
than Openshaw (most notably the Sokal affair).

The publication of *Ground Truth* (Pickles, 1995b) moved beyond an er
polemic and showed some attempt to include authors from inside GIS alom
critics. Despite the effort, the insiders stuck to their scientism (Goodchi
Goodchild, 1995, in particular), and the critics to theirs (Pickles, 1995a, ta
totalitarian tendencies). While this book was quite important, it did not serve to
a new common ground, but more to demonstrate how divergent the views wer
literature continued (Sheppard, 1995; Curry, 1998), the focus concentrated on th
(potential, imagined, observed) of GIS on society. Only a few talked about th
that our current GIS might be conditioned by societal pressures, cultural presup
and political choices (Chrisman, 1992, 1996).

Arguments about GIS technology often slip into a discourse of tech
determinacy. GIS-proponents and critics alike assert, consciously or unconscio
technology is intrinsically independent from the social world. This perpetuate
major tenets of technological determinism: (1) technology engages unilinear
from less to more advanced systems; and (2) technology is an imperative to wh
institutions and people must adapt (Bijker *et al.*, 1987; Woolgar, 1987; Bijker
1992; Feenberg, 1995). Technological determinism leads to the belief that the te
can be studied solely by itself, outside of the context of its construction or
consequence, 'implications' remain as the sole issue.

Proponents often acclaim geographic information technology as the means
more efficient and socially equitable decisions. These proponents hope to c
subjective issues and rationalize the process of establishing consensus, so that
can be made objectively (Dobson, 1983; Openshaw, 1991, 1993). Most of this
aligns itself with a 'March of Progress' metaphor, an attitude about history wi
utility to detect the choices and inconsistencies involved in technologica
(Chrisman, 1993). The idea of an automated geography implies that the tech
somehow independent of the people, operating on its own internal logic. Criti
are quite justified in calling attention to flaws in the proponents' claims.

The heralds of progress create the impression that improvement is inexo
assured. The GIS bandwagon suggests that jumping aboard is the way to
technology can fulfill every demand, and bring you the world. Dobson pl
technology on a clear rational path towards a better tomorrow, arguing that
become a *sine qua non* for geographic analysis and research ... the beginning
technological, scientific, and intellectual revolution" (Dobson, 1993, p. 431). Th
of *Ground Truth* made much of the claims of GIS proponents (Pickles, 1995c)
the advertising of GIS vendors (Goss, 1995; Roberts and Schein, 1995). 1
arrogant the claim, the better it seems to serve the critics.

The critics (*e.g.* Smith, 1992; Curry, 1998; Pickles, 1995b; Sheppard, 19
also focused on the impacts of technology. They often portray the technology
out of social control, something external to the social discourse. They use a s

sophisticated form of C.P. Snow's (1959) 'two cultures' argument, saying that technologists are not connected to the same literature and not engaged in the same bases of theory. The gap between two discourses does not mean that technology and technologists do not respond to their own versions of social forces. Both proponent and critic alike need to see where exactly the social comes into GIS. It may not be in the places they are watching.

Technological determinism, proclaimed by proponents or implied by critics, obscures the relationships between GIS technology and society largely by neglecting linkages. The contention between progress-believing technologists and humanistic-orientated social theorists omits the people involved with the technology and the complex interactions required to maintain it. GIS technology serves to extend human capabilities by other means, not a superorganic force in itself. The people who use GIS are not mere instruments of progress towards better information systems nor are they simply victims of its social consequences. The systems now in place reflect many layers of negotiation between social goals and technical capacity to respond. The simplistic metaphors must be replaced with more nuanced understanding of interactions between people and technology.

Rather than a vast superhuman realm, GIS technology is the result of localized social construction. This construction occurs when the technology is created, and continues as it is configured for each application. The march of progress myth must be replaced with a careful examination of the social divisions created and maintained by geographic information technologies. This paper will consider one example of these social divisions, after it presents some approaches to technology and society that move beyond technological determinism.

2.5 STUDIES OF SCIENCE AND TECHNOLOGY (STS)

In place of the technological determinism common in treating GIS, this paper draws specifically on recent theoretical insights from a number of interlocking literatures including the sociology of scientific knowledge (SSK), studies of technology and science (STS), history of technology and of science, philosophy of science and related fields. The twentieth century began with a fairly coherent expectation of the cumulative development of scientific knowledge (Carnap, 1966). By mid-century, the logical positivists seem to have conquered all opposition, broadcasting a message of method as a path of coherent science. Kuhn (1970) introduced an observation that science in this period was by no means as linear as it was meant to have been. The development of relativity in physics, for example, required replacing the whole 'paradigm', not just the incremental accumulation of adjustments to earlier schemes. Kuhn's approach left science (and thus technology) fairly independent from social concerns. Kuhn's work was so pervasive that the quantifiers in geography adopted the terminology of paradigms (*e.g.* Berry, 1973), a basically anti-positivist theory of knowledge. Some recent studies in the history of science (Galison, 1997) demonstrate further refinements in understanding how science operates, extending the concept of paradigms to allow for greater ambiguity in the negotiations between theorists and instrumentalists. The assurance that a particular scientific method always works has been strongly questioned (Feyerabend, 1993). Thus, the history and philosophy of science no longer provide support for the old mythology of inexorable progress.

Studies of science and technology (*e.g.* Barnes, 1974; Bloor, 1976; Latour & Woolgar, 1986) provide strong documentation of complex networks linking social organization, political structure, economic interaction, and cultural foundations to the development of a technology. The sociology of scientific knowledge developed a 'strong program' of researchers (Bloor, 1976; Collins, 1981) who argued that social relationships

underpin the development of science and technology. This strong program argue
the study of 'impacts' from technology to society. The constructivist literature (
Woolgar, 1986; Bijker *et al.* 1987; Latour, 1987, 1988; Bijker & Law, 199
though inherently quite diverse and far from unambiguous, modified the unid
direction providing a more complex dynamic of mutual constitution. Latour (199
that the division between 'nature' – a realm of scientific enquiry – and 'society'
for human creation – obscures intricate interactions required to sustain th
networks of current technology.

This literature argues that science and technology are constructed
multiplicity of viewpoints, and that this construction is distinctly local, not
Multiple social forces interact in the process of developing a complex technolog
GIS. Implementation of any technology depends on the specific local environ
strongly constrains how actors interact with the artifacts they construct. This
digs deeper than the argument of 'inherent logic'. Any logic in a technology was
by developers and adopted by users each group acting for their own reasons.

Social constructivist approaches provide a theoretical framework for exam
understanding the tight linkages between the actions of people and the techno
create and use. The web of technology and society consists of many
relationships between artifacts and people, institutions and data, software and res
Martin (2000) has applied this actor-network approach to demonstrate how diff
organizations interact in Ecuador. In his analysis, there are 'social' institutional l
in with 'technical' logics in a way that defies any simple explanation of one
overruling the other.

The STS literature offers some additional analytical approaches. Star, C
(1989) and Fujimura (1992) develop the relationships between multiple ac
artifacts through what they call *boundary objects*. Boundary objects mediate
different groups; they serve a dual function: at the same time they serve to di
differences, they also supply common points of reference (Harvey & Chrisma
Institutions and disciplines play a crucial role in formulating boundary objects t
for stable translations between different perspectives on the same phenomenon
(1997) provides a further development of these boundary concepts that may ap
directly to the interdisciplinary nature of GIS practice. He argues that translatio
too much mutual comprehension; he uses the linguistic metaphor of a pidgi
operating in a 'trading zone'. This concept offers an important insight for the
GIS technology, particularly in tempering the rush for a universal ontology
'common sense'.

2.6   CONFIGURING THE USER

As a specific resource for this paper, I am borrowing the phrase 'configure the u
a paper written by Steven Woolgar (1991), a British sociologist of techno
science. In the days when the IBM XT was the dominant PC, he observe
microcomputer manufacturing organization decided how to design their next m
argues that the group did not configure a machine to suit a specific body of u
rather that they built the machine that they could and attempted to configure the
suit the machine. He was contributing to a literature about the role of technica
(*e.g.* Latour & Woolgar, 1986, 1987). This theme has recently been extended (i
interactive form) to the study of software developments (Mackay *et al.,* 200
concept provides the most direct linkage to the critique of user-needs assessme
1987 paper. I only have space to introduce a few examples here.

Lets start out with the concept of 'user' in the first place. This term pollutes our understanding of the interaction between software, computer, data, organizations and the multiple skills of the varied people involved in any real-life GIS. Software vendors thrive by making these people dependent on the software, dependent on never switching packages. The annual user conference for dominant software vendors is lavish, and thoroughly suffused with a division of labor between the few software designers and the many users. The next version – just around the corner, real-soon-now – will clear up all problems, bring peace on Earth, and other minor side effects. The user is being configured in a social network of dependency, rather than treating the software as just one component, and often the least important compared to the data and the analytical models required to do any real work.

In an earlier paper written for a conference on error analysis (Chrisman, 1999), I examined a very specific GIS operation to register a digitized source to a coordinate system. This GIS operation adopts the mathematical formulation from photogrammetry that divides error into three parts: systematic, random and blunders. An appropriate numerical procedure, typically least-squares regression, can minimize the random error, and estimate the parameters of the systematic error (which can be totaly removed). This depends on having no 'blunders' to contaminate the estimation process. The choice of least-squares is not as straight-forward as it might seem. It is only optimal under some very special conditions, ones that are unlikely in routine use (Unwin & Mather, 1998). The technical details are important, but my point here is that the software designer isn't just choosing what mathematical model to apply, but who does what kind of work. The model handles two parts of the error, but not the other. The user is meant to weed out blunders, even though the software provides very limited tools to detect and remove them. If this analysis of errors were universal, then there would be little point in complaining. However, there are alternatives (from the realm of robust statistics) that divide error differently, and thus divide the labor along different lines. Such a change may not seem like much of a difference, but it uncovers a set of expectations about who is meant to know how much, which disciplines are best prepared to address certain issues, and so on. These are important elements in the present and future of GIS. If we continue to use least-squares, we are requiring users to perform a certain kind of blunder detection to protect the software's mathematics. Yes, we have software that perfomrs the 'optimal' solution, but only under unrealistic conditions. With the normal error-prone data entry of real-life users, least squares solutions are much worse than they need to be. If we convert to robust technology, the balance shifts. A small change in software technology configures the user in the sense of designing what a user has to know.

In a broader sense, the organization of data in a GIS serves to decide what kind of administrative arrangements belong in a GIS. The layer-cake design (that I did much to promote at Harvard and at Wisconsin) reflects a particular expectation of cooperating agencies each with their independent coverage. There were alternatives in our past, such as the integrated terrain unit techniques promoted by the Australians (Mabbutt, 1968) and adopted by the Food and Agriculture Organization of the UN (FAO, 1976). The layer based design is not neutral, it makes certain kinds of arms-length custodian approaches easier. While I can contend that these are more likely to succeed and perhaps more likely to contribute to equity, such a decision should not be furtive. Rather than configuring the user, we do need to ensure that the software is not out of control.

One possible path towards such change in control is set forth as the choice between the paradigms of the Cathedral and the Bazaar (Raymond, 1998). Raymond sets out a distinction between centralized software development (the cathedral) and a democratized, decentralized scheme (the bazaar). Certainly, most commercial GIS suffers from the divisions of labor implicit in the cathedral model. Broader participation and freer

dissemination might unblock some of these barriers, also changing the
configuring the user. I do accept some of Bezroukov's (1999) critique of Ray
distinctions are not quite as easy. In particular, the role of social status in c
networks limits the pure dissemination of a bazaar. In any even, this calls
examination of the social roles in GIS software.

2.7 CONCLUSION

The 1987 paper argued that the fundamental principles of GIS were to be fou
social goals that a GIS serves. While this may still be a grand objective, the
seems a bit more complex in the light of understanding the criticism of GIS fr
and outside. Society and GIS are no longer simple categories that remain
Software and databases are integrally implicated in social relationships at all le
should be no surprise, rather it should focus each of us on communicating all th
to all affected parties. The social goals may not need consideration just at the ou
each step and each detail of GIS operation.

2.8 ACKNOWLEDGEMENTS

Partially supported by a grant from National Science Foundation (USA) Grant
10075. AGI provided funding to attend and present this paper at GISRUK 2001

2.9 REFERENCES

Barnes, B., 1974, *Scientific Knowledge and Sociological Theory*. (London: Rot
Kegan Paul).

Berry, B.J.L., 1973, Paradigms for modern geography. In *Directions in Ge*
edited by Chorley, R.J., (London: Methuen), pp. 3-22.

Bezroukov, N., 1999, A second look at the cathedral and the bazaar. http
monday.dk/issues/issue4_12/bezroukov/index.html.

Bie, S., 1984, Organizational needs for technological advancement. *Cartograph*
pp. 44-50.

Bijker, W.E., Hughes, T.P. and Finch, T.J. (editors), 1987, *The Social Constr*
Technological Systems: New Directions in the Sociology and History of Te
(Cambridge: MIT Press).

Bijker, W.E. and Law, J. (editors), 1992, *Shaping Technology/Building*
(Cambridge: MIT Press).

Bloor, D., 1976, *Knowledge and Social Imagery*. (London: Routledge & Kegan I

Burrough, P.A., 1986, *Principles of Geographical Information Systems j*
Resource Assessment. (Oxford: Clarendon Press).

Carnap, R., 1966, *Philosophical foundations of physics; an introductio*
philosophy of science. (New York: Basic Books).

Chrisman, N.R., 1975, Topological data structures for geographic representation
the 2nd *Int. Symp. on Computer-Assisted Cartography (Auto–Carto II)*, pp. 34

Chrisman, N.R., 1987a, Fundamental principles of geographic information sy
Proc. of the 8th Int. Symp. on Computer-Assisted Cartography (Auto (
Baltimore, edited by Chrisman, N.R., pp. 32-41.

Chrisman, N.R., 1987b, Design of information systems based on social and cultu

Photogrammetric Engineering and Remote Sensing **53**, pp. 1367–1370.

Chrisman, N.R., 1992, Ethics for the practitioners of Geographic Information Systems embedded in 'Real World' constraints of guilds, professions and institutional sponsorship. *Proceedings GIS/LIS 92* **1**, pp. 129–137.

Chrisman, N.R., 1993, Beyond spatio-temporal data models: a model of GIS as a technology embedded in historical context. *Proceedings Auto-Carto II*, pp. 23–32.

Chrisman, N.R., 1996, GIS as social practice. In *Technical Report 96-7*, edited by Harris, T. and Weiner, D., pp. D9–D11. (Santa Barbara: National Center for Geographic Information and Analysis).

Chrisman, N.R., 1999, Speaking truth to power: An agenda for change. In *Proc. 3rd Int. Symp. on Accuracy of Spatial Databases Applied to Natural Resource Issues*, edited by Lowell, K. and Jaton, A. (Ann Arbor: Ann Arbor Press).

Collins, H.M., 1981, Understanding science. *Fundamentals of Science* **2**, pp. 367–380.

Curry, M.R., 1998, *Digital Places: Living with Geographic Information Technologies.* (New York: Routledge).

Dobson, J., 1983, Automated geography. *Professional Geographer* **35**, pp. 135–143.

Dobson, J., 1993, The geographic revolution: A retrospective on the age of automated geography. *Professional Geographer* **45**, pp. 431–439.

FAO, 1976, *A Framework for Land Evaluation.* (Rome: Food & Agriculture Organization).

Feenberg, A., 1995, Subversive Rationalization: Technology, Power, and Democracy. In *Technology and the Politics of Knowledge*, edited by Feenberg, A. and Hannay, A. pp. 3–22. (Bloomington: Indiana University Press).

Feyerabend, P.A., 1993, *Against method: outline of an anarchistic theory of knowledge.* (London: Verso) third edition.

Fujimura, J.H., 1992, Crafting science: standardized packages, boundary objects, and 'translation'. In *Science as Practice and Culture*, edited by Pickering, A. pp. 168–211. (Chicago: University of Chicago Press).

Galison, P., 1997, *Image and Logic.* (Chicago: Chicago University Press).

Goodchild, M.F., 1992, Geographical information science. *International Journal of Geographical Information Systems* **6**(1), pp. 31–45.

Goodchild, M.F., 1995, GIS and geographic research. In *Ground Truth: The Social Implications of Geographic Information Systems*, edited by Pickles, J. pp. 31–50. (New York: Guilford).

Goss, J., 1995, Marketing the new marketing: the strategic discourse of geodemographic information systems. In *Ground Truth: Social Implications of Geographic Information Systems*, edoted by Pickles, J. pp. 130–170. (New York: Guilford).

Harley, J.B., 1989, Deconstructing the map. *Cartographica* **26**(2), pp. 1–20.

Harvey, F. and Chrisman, N.R., 1998, Boundary objects and the social construction of GIS technology. *Environment and Planning A* **30**, pp. 1683–1694.

Herring, J., 1987, TIGRIS: topologically integrated geographic information system. In *Proc. of the 8th Int. Symp. on Computer-Assisted Cartography (Auto Carto 8)*, Baltimore, edited by Chrisman, N.R., pp. 282–291.

Jordan, T., 1988, President's column: the intellectual core. *Newsletter of the Association of American Geographers* **23**(5), pp. 1.

Kuhn, T.S., 1970, *The Structure of Scientific Revolutions.* (Chicago: University of Chicago Press) second edition.

Latour, B., 1987, *Science in Action.* (Cambridge: Harvard University Press).

Latour, B., 1988, *The Pasteurization of France.* (Cambridge: Harvard University Press).

Latour, B., 1993, *We Never Were Modern.* (Cambridge: Harvard University Press).

Latour, B. and Woolgar, S., 1986, *Laboratory Life: The Construction of Scientific Facts.* (Princeton: Princeton University Press) 2nd edition.

Mabbutt, J.A., 1968, Review of concepts of land classification. In *Land E*
edited by Stewart, G.A. pp. 11–28. (Melbourne: Macmillan).

Mackay, H., Carne, C., Beynon-Davies, P. and Tudhope, D., 2000, Reconfig
user: using rapid application development filing cabinet. *Social Studies of Sc*
pp. 737–757.

Martin, E.W., 2000, Actor-networks and implementation: examples from cor
GIS in Ecuador. *International Journal of Geographical Information Scienc*
pp. 715–738.

National Research Council, 1983, *Procedures and Standards for a Mul*
Cadastre. Panel on a Multipurpose Cadastre. (Washington DC: National
Press).

Openshaw, S., 1991, A view on the crisis in geography, or using GIS to put
Dumpty back together again. *Environment and Planning A* **23**, pp. 621–628.

Pickles, J., 1993, Discourse on method and the history of a discipline: refle
Jerome Dobson's 1993 'Automated Geography'. *Professional Geographe*
451–455.

Pickles, J., 1995a, Conclusion: Toward an ecomony of electronic representatio
virtual sign. In *Ground Truth: Social Implications of Geographic Inf*
Systems, edited by Pickles, J., pp. 223–240. (New York: Guilford).

Pickles, J. (editor), 1995b, *Ground Truth: Social Implications of Geographic Inf*
Systems. (New York: Guilford).

Pickles, J., 1995c, Representations in an electronic age: geography, GIS and de
In *Ground Truth: Social Implications of Geographic Information Systems,*
Pickles, J., pp. 1–30. (New York: Guilford).

Raymond, E.S., 1998, The Cathedral and the Bazaar. http://firstmonday.dk,
issue3_3/raymond/index.html.

Roberts, S.M. and Schein, R.H., 1995, Earth shattering: global imagery and
Ground Truth: Social Implications of Geographic Information Systems,
Pickles, J., pp. 171–195. (New York: Guilford).

Sheppard, E., 1995, GIS and society: towards a research agenda. *Cartogra*
Geographic Information Systems **22**, pp. 5–16.

Smith, N., 1992, History and philosophy of geography: real wars, theory wars.
in Human Geography **16**, pp. 257–271.

Snow, C.P., 1959, *The Two Cultures and the Scientific Revolution.* (Ca
Cambridge University Press).

Star, S.L. and Greisemer, J.R., 1989, Institutional ecology, 'translations', and
objects: amateurs and professionals in Berkeley's Museum of Vertebrate
Social Studies of Science **19**, pp. 387–420.

Taylor, P. and Overton, M., 1991, Commentary: further thoughts on geography
Environment and Planning A **23**, pp. 1087–1094.

Unwin, D.J. and Mather, P.M., 1998, Selecting and using ground control points
rectification and registration. *Geographical Systems* **5**(3), pp. 239–260.

Woolgar, S., 1987, Reconstructing man and machine: a note on sociological cri
cognitivism. In *The Social Construction of Technological Systems: New Dire*
the Sociology and History of Technology, edited by Bijker, W.E., Hughes
Finch, T.J., pp. 311–328. (Cambridge: MIT Press).

Woolgar, S., 1991, Configuring the user: the case of usability trials. In *A Soc*
Monsters: Essays on Power, Technology and Domination, edited by Law, J.
97. (London: Routledge).

PART I

GIS and Crime

3

When is a hotspot a hotspot? A procedure for creating statistically robust hotspot maps of crime

Spencer Chainey, Svein Reid and Neil Stuart

3.1 INTRODUCTION

Several techniques and algorithms are used in practice for the generation of continuous surface hotspot maps, all of which have different merits. These mainly relate to their ease of use, application to different types of events, visual results and interpretation. Few of these methods help to distinguish a consistent defining threshold that helps the analyst decide when a cluster of crimes can be defined as a 'hotspot'. This paper will present results of some partnered research that explores a procedure for creating statistically robust hotspot maps. Our methods include the application of point pattern analysis techniques to identify for spatial clustering, spatial dispersion, spatial autocorrelation, and Local Indicators of Spatial Association (LISA statistics), plus review the use of spatial analysis tools to visualise and assist in formatting the design of the crime hotspot map and the definition of hotspot thematic thresholds.

Maps showing the distribution of crime as a digital continuous surface are increasingly replacing point (or pin) maps, thematic boundary maps and spatial ellipses as ways to visualise and understand patterns of crime events. Each of these techniques has their application but suffer from problems that relate to their operation and interpretation of crime event patterns (see Table 3.1).

The increased application of continuous surface smoothing methods has been in partial recognition to these problems, but is also linked to their more common availability, their appealing visual allure, and the improved precision and accuracy of geocoded crime records. The creation of these continuous surface or 'hotspot maps' allow for easier interpretations of clusters of crime, reflect more accurately the spatial distribution of the crime hotspot, and often have fewer parameters to set. Continuous surface smoothing methods also consider concentrations of crime at all event levels, rather than cluster grouping some and discounting others. However, as their appeal has increased few questions are being made of the outputs generated. Instead, many agencies that use these methods are becoming easily caught in the 'false lure' of the sophisticated

looking geo-graphic they have produced, being reluctant to question its valid
statistical robustness. Where this is particularly the case is when little regard i
the legend thresholds that are set that help the analyst decide when a cluster of c
be defined as a hotspot. This visual definition of a hotspot being very much
'whims and fancies' of the map designer. For example, a map showing the di
of crime as a continuous surface can have as little or as many hotspots on it dep
the ranges selected by the map designer to show spatial concentrations of th
events. Different methods also produce different surface results (Jefferis, 1999
often further mislead.

Table 3.1 Crime mapping applications and problems associated with point patterns, thematic boun
and spatial ellipse methods.

Mapping Method	**Mapping Application**	**Problems of Method**
Point Patterns	• Individual incidents. • Small volumes of crime. • Repeat locations (by graduating symbol size).	• Inappropriate as meth interpret crime cluste (Jefferis, 1999, Raym 1995).
Thematic Boundary Maps	• Reports of crime events by administrative area. • Integration of crime data with other boundary associated data (*e.g.* calculate burglary rates or perform area correlations with Census data).	• Modifiable area unit (see Bailey and Gatre 1995, or Openshaw, • Restricted to visualis events by boundaries than as a continuous gradient.
Spatial Ellipses	• Grouping of crime clusters. • Revealing areas for closer inspection.	• Do not represent the a spatial crime distribut crime hotspots do not naturally form spatial ellipses). • The need to enter a la number of parameter influence variability i final output). • By grouping events i elliptical clusters, exc from any visual resul events that do not bel cluster.

This paper suggests a pragmatic procedure for creating statistically robust hotspot maps of crime. The procedure includes provisional steps that test for clustering and dispersion, reviews three methods (kernel density estimation, inverse distance weighting, and a location profiler) that have previously been applied to convert point maps of crime into crime pattern surfaces, explores the ways to standardise parameter settings of these continuous surface smoothing methods, and suggests an easy to apply and interpret method for defining crime hotspot thresholds. We also demonstrate options to extend the procedure to a more advanced level of analysis using local indicators of spatial association (LISA statistics).

The data we use in our tests are property/location precise allegations of crime from the Metropolitan Police Crime Report Information System for the London Borough of Hackney. We experiment with four different categories of crime – all crime, robbery, residential burglary, and vehicle crime – for the three-month period, June to August 1999.

3.2   TESTING FOR CLUSTERING AND DISPERSION

A primary step that is often avoided but is fundamental to the detection of clusters of point events are global tests that indicate if clustering and dispersion exist in the original point distribution. This first step can provide insight into what types of patterns will be expected when the mapped crime points are converted into a continuous surface. For example, whether the points show evidence of clustering or are randomly distributed, and how dispersed the distribution is relative to other crime types. Crime analysts often assume crime distributions to be clustered in some form or other, and whether clusters do or do not exist, some can be identified from random crime distributions.

There are several approaches for analysing a point distribution for spatial randomness. Most of them incorporate the basic principles of hypothesis testing and classical statistics, where the initial assumption is that the point distribution is one of complete spatial randomness (CSR). By setting the CSR assumption as the null hypothesis the point distribution can be compared against a set significance level to accept or reject the null hypothesis. Spatial autocorrelation tests have been used previously to test for evidence of crime event clustering (Chakravorty, 1995). Spatial autocorrelation techniques do though require an intensity value, beit a weighting linked to the event or a count of crimes where the crime point relates to the centroid of an area to which crime events have been aggregated. From our point of view, if the original crime event data exists as accurate point georeferenced data, aggregating this data to a common point will lose spatial detail. With the increased availability of accurate and precise geocoded records of crime it would seem more important to use methods that do not require an intensity value but that retain and perform tests on the original crime event point data. *Point retaining methods* only require x (eastings) and y (northings) coordinates which are used to test for evidence of point clustering and dispersion.

Where the user has access to data where each point relates to individual crime events (irrespective of whether some of these events are mapped to exactly the same location) we suggest the point retaining methods of nearest neighbour analysis (NNA) to test for clustering and standard distance for dispersion. Both these tests are familiar to many crime analysts, being available in STAC (Illinois Criminal Justice Information Authority, 1996) and CrimeStat (United States Institute of Justice, 1999). They are also simple and quick to apply and provide a good global measure that can help lead the analyst in creating more confident mapping output.

If analysts only have access to crime point data that are aggrega (representing the number of crime events within a certain geographic area, *e.* tracts) we suggest the use of the spatial autocorrelation technique, Moran's I, clustering. The statistic works by comparing the value at any one location with at all other locations (Levine, 2002, Bailey and Gatrell, 1995, Anselin, 199 1985). The significance of the result can then be tested against a theoretical d (one that is normally distributed) by dividing by its theoretical standard dev more details on this test see Levine, 2002).

NNA and standard distance tests were performed on the crime data availa London Borough of Hackney. Table 3.2 shows that all crime data point se evidence of clustering in their distribution. The degree of difference in the clustering between the two different bounding areas was noted with additi suggesting the need to base all calculations on the true boundary area (Reid, 199

Table 3.2 Standard distance and nearest neighbour analysis results for four categories of crime in Borough of Hackney (June 1999 – August 1999). All categories show evidence of crime point ever and describe relative differences in dispersion.

Crime Type and Bounding Area	Standard Distance	NN Index	z-score	Descriptic Crime Distr
All Crime				
Bounding rectangle area		0.32	-129.11	Cluster
True boundary area	1807.94 m	0.46	-103.14	Cluster
Robbery				
Bounding rectangle area		0.59	-19.14	Cluster
True boundary area	1749.94 m	0.80	-9.20	Cluster
Residential Burglary				
Bounding rectangle area		0.57	-27.14	Cluster
True boundary area	1806.28 m	0.74	-16.46	Cluster
Vehicle Crime				
Bounding rectangle area		0.52	-38.73	Cluster
True boundary area	1820.85 m	0.72	-22.16	Cluster

To test for dispersion we calculated each crime category distributions' distance. This measure could then be used on a relative scale to describe differ of dispersion between crime types. The lower the relative standard distance dispersed the crime distribution. Robbery was seen to be the least dispersed cr suggesting a distribution that is concentrated (clustered) at a few locations. R burglary showed evidence of dispersion similar to all crime, whilst vehic although being clustered, tended to be more dispersed.

We now have evidence of clustering for all crime types and an indicatc relative levels of dispersal. A final preparatory step before creating hotspot crime was to carry out nearest neighbour tests for each point's 20 nearest neigh a K order of 20). This would return mean nearest neighbour distances for 1, 2 nearest neighbours that could be used to help create search radii for the continuous surface smoothing mapping methods (Williamson *et al.*, 1999). By these simple global tests for clustering and dispersal we can more confidently p creating hotspot maps of crime.

3.3   CONTINUOUS SURFACE SMOOTHING METHODS

The methods we reviewed were those that were available through commonly used geographical information systems software packages. This review allowed us to re-evaluate techniques that the London Borough of Hackney had employed in creating hotspot maps of crime and decide on a method that we could focus on for further analysis. The methods we considered were kernel density estimation, location profiler (unique to the MapInfo add-on, Vertical Mapper) and inverse distance weighting.

3.3.1 Kernel Density Estimation

The kernel density method creates a smooth surface of the variation in the density of point events across an area. The method is explained in the following steps:

- a fine grid is generated over the point distribution (see Figure 3.1);

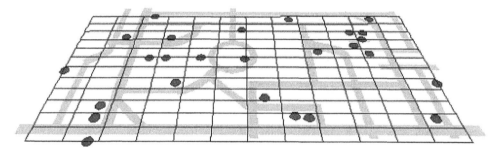

Figure 3.1 Step 1 kernel density estimation – a fine grid is placed over the area covered by the crime points. In most cases the user has the option to specify the grid cell size. (Source: Ratcliffe, 1999a.)

- a moving three-dimensional function of a specified radius visits each cell and calculates weights for each point within the kernel's radius. Points closer to the centre will receive a higher weight, and therefore contribute more to the cells total density value (see Figure 3.2);

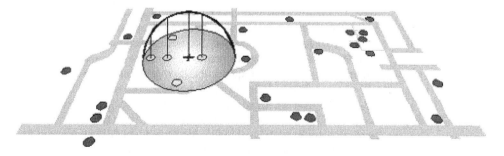

Figure 3.2 A search radius (or bandwidth) is then selected, within which intensity values for each point are calculated. Points are weighted, where incidents closer to the centre contribute a higher value to the cells intensity value. (Source: Ratcliffe, 1999a.)

- final grid cell values are calculated by summing the values of all circle su each location (see Figure 3.3).

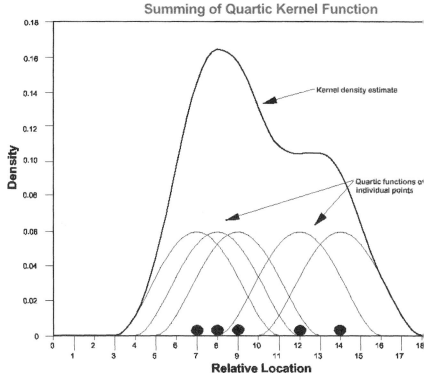

Figure 3.3 Each grid cell value is the sum of values of all circle surfaces for each locatio
(Source: Levine, 2002.)

The quartic kernel estimation method we applied requires two parameters prior to running. These are the grid cell size and bandwidth (search radius). bandwidth is the parameter that will lead to most differences in output when va method for choosing bandwidth that we have adopted follows from that of Will *al.* (1999) where bandwidth relates to the mean nearest neighbour distance for orders of K. This still requires the user to make a choice over which K order We regard this as an advantage of the technique where users are prompted to th the data they are mapping and apply the K order's mean nearest neighbour dista bandwidth relevant to their application. For example, if the application is requires output for focused police patrolling a smaller bandwidth would be usec that requires a more strategic view of crime hotspots. Having this flexibilit demonstrated in Section 3.4.

Users are also required to specify a grid cell size. We again regard this as feature in kernel estimation whereupon users have the flexibility to set sizes r the scale at which the output will be viewed. Large cell sizes will result in coarser or blocky looking maps but are fine for large scale output, whilst smaller cell sizes add to the visual appeal of the continuous surface produced but may create large file sizes. Where the user is unsure over the cell size to use, we suggest following the methodology of Ratcliffe (1999b) where cell size resolution is the result of dividing the shorter side of the minimum bounding rectangle by 150. Cell size for the London Borough of Hackney was set to 25m. The cell values generated from quartic kernel estimation are also in meaningful units for describing crime distributions (*e.g.* the number of crimes per square kilometre).

3.3.2 The Location Profiler

The Location Profiler method uses a technique where the distance from an overlaid grid cell to a specified number of nearest points within a specified radius are measured, these distances are then summed and the resulting value used to indicate local intensities (Northwood Geoscience, 1998). The first step therefore requires a grid of user defined size to be overlaid across the point distribution. Size of the grid cell can follow the flexible methodology to set this parameter as previously described. Distances from the centre of each cell to a specified number of points, or points within a specified search radius are then measured. Choice of the number of points to measure distances to is somewhat arbitrary and as Eck *et al.* (1998) note, there exists no underpinning theory to aid its choice. Search radius can follow from the methodology described for choosing kernel bandwidth, where the radius entered relates to a mean nearest neighbour distance for a particular K order.

A problem with the Location Profiler is that cell values generated do not fit logically as outputs expected to describe crime hotspots. The cell values refer to the average distance to points within the specified search radius. Cells with low values represent areas of high crime density, and cells with high values are areas of low density. These values can be visualised thematically on a map, but the cell unit values can add confusion when attempting to explain the differences between local patterns of crime.

By choosing a search radius based on the mean nearest neighbour distances approach, the effect was to highlight isolated crime events rather than highlighting areas of obvious clustering. This occurred because there were no other points within the search radius to measure distances to, the average distance to all other points being zero. Average distances in areas of crime clustering would be low, but greater than for those areas where there are these isolated points (see Figure 3.4).

A better approach in using Location Profiler would be to set the search radius to cover the whole point distribution, set the display radius as a K order mean nearest neighbour distance and choose an appropriate number of points to measure distances between. Choosing an appropriate number of points is still an arbitrary process. If we were to base the number of points on the K order of the mean nearest neighbour distance used as the display radius (*e.g.* if a K order of 2 returned a mean nearest neighbour distance of 216m, 216m would be the display radius and 2 would be the number of points), in cases where our point number was small, areas of isolated points would often appear as being just as significant crime hotspots than areas where there was an obvious cluster of crimes (see Figure 3.5).

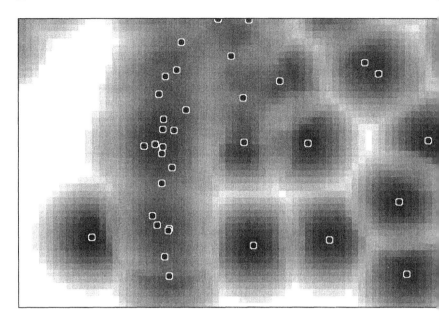

Figure 3.4 Location Profiler returns cell output values based on the average distance from a num▮ within a specified radius. In the figure above, the darker the shading the lower the average d▮ choosing a radius based on descriptions of the crime points spatial distribution (*i.e.* K order n▮ neighbour distances), the effect was to highlight isolated crime events where there were no other ▮ the search radius, rather than identifying areas of obvious clustering.

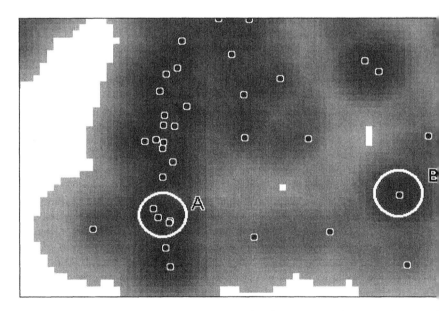

Figure 3.5 Location Profiler parameters were set as the whole study area for the search radius, 2▮ display radius (the mean nearest neighbour distance for 2 orders based on the point distribution) and of points to visit to measure distances as 2 (the K order). The figure shows that if we use small points, areas that are relatively isolated but that have two points at the same location (area highlighted at least (and often more) significantly than areas of obvious greater clustering (area A).

3.3.3 Inverse Distance Weighting

The inverse distance weighting method (IDW) uses a floating circle to visit each new node of a grid created across the distribution of points, assigns weights to the data points lying within a prescribed distance, which are then averaged to return a value for each grid cell. The weights assigned to points are based on their distance from the grid cell node. The IDW method is most commonly applied to data which can be extremely variable over short distances (Northwood Geoscience, 1998). For this reason the method has been used as a technique to help visualise spatial patterns of crime. However, in its strictest sense IDW is an interpolation technique that requires an intensity value for each crime event. Interpolation techniques are most often used to generate estimates of an attribute at unsampled locations. For example, interpolation methods can be applied to create continuous surface maps of rainfall distribution, where the point locations are rain gauges and the attribute values attached to them is the volume of rainfall collected in the gauge. When we create continuous surface maps of crime we are not estimating the number of crimes at 'unsampled' locations, but instead using the point distribution to perform an area data aggregate which can be smoothed through application of a suitable spatial function to describe crime pattern distribution.

To apply IDW an intensity value is required. For each crime point this intensity value can be set to '1', such that each point has equal intensity. Parameters required for IDW include a search radius, display radius, cell size and exponent. There is little guide as to what exponent to set, but experience in Hackney has always left this setting as the default. Grid cell size again is best set in relation to the scale of viewed output, whilst search and display radii can follow the mean nearest neighbour distance settings previously described. IDW does though require an initial point aggregation technique to be applied. The resulting output of this required technique will mean the loss of some spatial detail, therefore questioning already the detailed accuracy of the final continuous surface output. A *coincident point distance* parameter setting is also required to apply data aggregation. This coincident point distance is a search radius applied to each new grid node across a grid network covering the distribution of points. In this sense it is similar to the location from which a search radius or bandwidth is applied in the Location Profiler technique and the quartic kernel density estimation technique.

Through its nature, the IDW method does not honour local high or local low values. As Figure 3.6 shows, the averaging method will create output values that are 'cooled' at local high points, and that could be over emphasised in low value areas.

The cell values returned also relate to the values of nearby aggregated points and not to the original spatial point detail (see Figure 3.7). The moving window average technique can also lead to visual inaccuracies relating to crime distribution. An example of this is shown in Figure 3.7 where the region to the west of the main hotspot shows a small increase in crime intensity, even though there are no crimes in that area to suggest this increased intensity. Description of the cell values returned can be best made by rounding up the values to integers and using these integers with the search radius to describe local densities. For example, at the hotspot indicated by the darkest area, the distribution of crime relates to the number of points (*i.e.* the cell value, 5) within 216m (the search radius) of the point of interest. This description can though lead to confusion where there is high variability across short distances.

From this review it was decided that the quartic kernel density estimation method provided the best opportunity to proceed with creating statistical robust crime hotspot mapping output. The method creates understandable grid cell value outputs that relate to

crime, plus requires fewer parameters to be set than those required for m
methods and where the parameters entered can relate to the spatial distribut
points being analysed. The method also has the advantage of deriving crin
estimates based on calculations performed at all locations (Levine, 2002), a
some practical flexibility in map design.

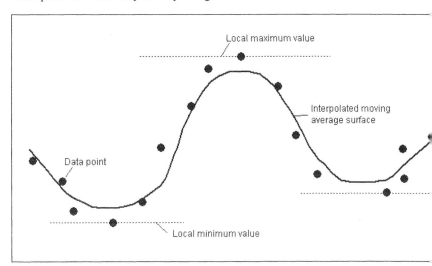

Figure 3.6 The surface created using the moving average technique in the IDW method is less th
max. value and greater than the local min. value. (Source: modified from Northwood Geoscience, 1

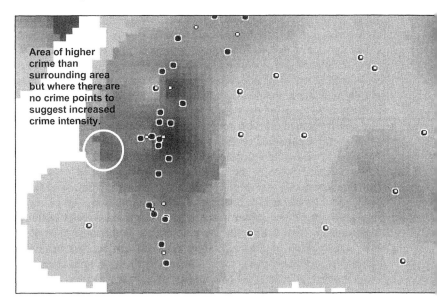

Figure 3.7 Inverse distance weighting output using a search radius of 216m (this equals the K o
nearest neighbour distance of the point distribution). To apply IDW, a data aggregation routine in
to be carried out. Data aggregation was applied with a coincident point distance search radius of
data aggregation output is displayed above as the lighter points. The moving average technique a
inverse distance weighting method has the tendency to 'cool' values at local high points, and can l
inaccuracies of crime distribution.

3.4   DEFINING CRIME HOTSPOT LEGEND THRESHOLDS

In our aims to define crime hotspot legend thresholds we set out criteria that we felt were important in helping to decide on suitable method implementation. These criteria were:

- the method needed to be practical and simple to apply;
- the thresholds generated were meaningful and could be linked to a value that could be easily understood as a unit describing crime concentration;
- the separation of thematic thresholds followed a consistent methodology and where the upper most categories consistently defined when a crime concentration reached hotspot status;
- the method was more robust in taking into account the statistical spatial distribution of the point data;
- but still retained flexibility in map design, appropriate to output required at different scales and for different applications.

There is still much debate around the theoretical definition of understanding when a crime hotspot becomes a hotspot. By 'hotspot', we refer to areas of distinctly high concentrations of crime, relative to the overall distribution of crime across the borough (see also Home Office, 2001). This would mean that if clusters were identified from the nearest neighbour analysis tests, a consistent methodology would identify where these hotspots were located and indicate a level of concentration relative to the full crime distribution. By investigating crime hotspots on a relative scale means that we are not restricted to a static setting that defines levels in crime concentration. Implementing this relative approach would also enable analysts operating at different geographic scales (*e.g.* local, regional and national levels) to apply the same methodology to identify and prioritise the tackling of crime hotspots. We present results from two different techniques we have investigated for defining thematic hotspot legend thresholds.

3.4.1 Incremental Standard Deviation Approach

Our first approach for defining crime hotspot thresholds was to explore the use of incremental standard deviations above the grid cell mean. Standard deviations are useful in defining where in a distribution the cut off is between values of different statistical significance. The procedure we explored applied calculations on grid cells that had a value of greater than 0 and were within the study area boundary. From this grid cell set, the mean and standard deviation was calculated, with grid cell thematic thresholds set at:

- 0 to mean
- Mean to + 1 standard deviation (SD)
- + 1 SD to + 2 SD
- + 2 SD to + 3SD
- + 3 SD to + 4 SD
- Greater than + 4 SD

Plate 1 shows the application of this incremental standard deviation threshold approach for robbery (June to August 1999). Visually, the method is appealing, clearly identifying in a structured manner areas of increasing crime concentration significance. The method

is easy to apply, makes use of K order mean nearest neighbour distances
bandwidths, retains flexibility in the map design by allowing the user to apply d
order bandwidths and grid cell sizes to the suited application, and uses a
methodology to separate thematic thresholds. However, standard deviations a
easiest of statistics to describe to the novice crime hotspot map reader. If there
to explain the concept of standard deviations for identifying thresholds of si
before the crime hotspot map can be interpreted then the threshold meth
acceptable as it itself creates a barrier to map interpretation. Also, grid cell val
quartic kernel density output are not normally distributed. As our aim is t
robustness in our crime hotspot mapping output, it would seem inappropriate t
standard deviation approach to defining legend thresholds.

3.4.2 Incremental Mean Approach

Our second method applied incremental multiples of the grid cells' mean to de
hotspot map thresholds. Again, calculations were only applied to grid cells t
value of greater than 0 and were within the study area boundary. From this gr
the mean was calculated, with grid cell thematic thresholds set at:

- 0 to mean
- Mean to 2 mean
- 2 mean to 3 mean
- 3 mean to 4 mean
- mean to 5 mean
- Greater than 5 mean

Plate 2 shows the application of this incremental mean threshold app
robbery (June to August 1999). Comparisons in the map can be immediately d
the incremental standard deviation approach where the method is visually appe
structures the thematic thresholds to clearly identify areas of highest crime conc
The method is simple to apply, requiring only the calculation of the grid c
makes use of K order mean nearest neighbour distances to define bandwidth
flexibility in the map design by allowing the user to apply different K order b
and grid cell sizes to the suited application, and uses a consistent methodology t
thematic thresholds.

As a statistic, the mean is an easier value for the novice map reader
Increments of the mean would be more obviously linked to increasing values,
relative significance. This makes this approach immediately more appealing as
to define crime hotspot legend thresholds.

The incremental mean approach therefore shows potential in being a
method for defining crime hotspot legend thresholds. At this stage we appli
tests using this method but on different crime types and for different time peri
would allow us to investigate consistency in the technique, and explore the fle
map design, described as an advantage in the quartic kernel density estimation m

The maps in Plate 4 show the results of applying this incremental me
hotspot threshold approach to the four different types of crime and for
bandwidths. The incremental mean legend threshold approach consistentl
mapped crime hotspots. By using different bandwidths, different scales in outp

generated to suite different applications. Maps with small bandwidths demonstrate the spiky and more focused crime mapping output returned, whilst larger bandwidths tend to smooth the data and indicate, for example, general areas in need of strategic priority. As bandwidth is linked to the underlying spatial distribution of the crime point data, all the maps can be described as statistically robust, and use a consistent, easy to apply and easy to understand method for defining crime hotspot thresholds. In addition, we can now observe the global descriptions calculated in Section 3.2 and see that they match the crime hotspot mapping output. All maps display areas of crime clustering, and when compared against each other show relative levels of dispersal. For example, the hotspot maps of vehicle crime show a greater degree of dispersion compared to that of robbery.

3.5 ADVANCED HOTSPOT ANALYSIS: APPLICATION OF LOCAL INDICATORS OF SPATIAL ASSOCIATION

Local indicators of spatial association (LISA statistics) have been described as being particularly suited to identifying crime hotspots (Anselin, 1995; Getis and Ord, 1996). LISA statistics assess the local association between data by comparing local averages to global averages. For this reason they are useful in adding definition to crime hotspots, and placing a spatial limit on these areas of highest crime event concentration (Ratcliffe and McCullagh, 1998). One of the more applied LISA statistics on crime point events is the Gi* statistic (see Ratcliffe and McCullagh, 1998 for more details). The Gi* statistic is applied to a grid cell output, such as a quartic kernel density estimation map, from which local associations are compared against the global average. The user is typically required to enter a search distance within which local associations are explored, and a significance level related to the confidence in the final output. The search distance is usually set to three times the grid cell size of the original kernel estimation surface (*i.e.* for our data the search distance was set to 75m) and whilst levels in significance can often range between 99.9%, 99% and 95%, this range has far less effect over eventual hotspot areas than parameters set for grid cell size or bandwidth (Ratcliffe, 1999b). The most common significance level to apply is 99.9%.

Plate 3 shows the result of the Gi* LISA statistic for robbery. This LISA output has been mapped with the quartic kernel density estimation surface generated using a bandwidth of 118m (K2). The map shows the matching between this LISA output and kernel value results above the three mean threshold. This LISA output adds definition to our quartic kernel density estimation, indicating the level at which hotspots can be more clearly distinguished against other levels of crime concentration.

3.6 DISCUSSION

We can now suggest a procedure for creating statistically robust hotspot maps of crime:

- Apply nearest neighbour analysis to test for clustering
- Calculate standard distance to return a relative measure of dispersion
- Perform a nearest neighbour test for each point's 20 nearest neighbours (*i.e.* a K order of 20). Performing this test for at least each point's 20 nearest neighbours will return sufficient information to help set parameters for continuous surface smoothing

- Create a continuous surface map by applying the quartic kernel density ⁣
  method. Parameters should be set with respect to different K order me⁣
  neighbour distances, where the choice of K order is dependent on the crim⁣
  application
- Use an incremental mean approach to set grid cell thematic legend threshol⁣
- Advanced option – perform a Gi* LISA statistic analysis to quartic kern⁣
  This mapped LISA output will add definition to our current quartic kern⁣
  indicating the level at which hotspots can be more clearly distinguished ag⁣
  levels of crime concentration.

But when is a hotspot a hotspot? We argue that there is the need for fle⁣
defining when a crime concentration reaches hotspot status, depending large⁣
level of significance required in the result. However, from our results we su⁣
hotspot status is reached at 3 multiples of the mean grid cells' value and abc
defining threshold visually fits with a LISA statistic output of 99.9% significanc

Our methodology has yet to experiment with secondary variables that
environmental heterogeneity. This is where we would use a secondary v⁣
describe the underlying population (*e.g.* distribution of residential properti
which a continuous surface hotspot map would be an estimate of cr⁣
Theoretically, the incremental mean approach for defining hotspot thresholds ⁣
be successfully applied.

The crime hotspot maps we generate are also only snapshots of a particu⁣
in time. Future areas of work we aim to explore include space-time interaction,
statistics devised by Knox and Mantel (see Bailey and Gatrell, 1995). This wo⁣
whether certain types of crime display temporal hotspots in particular areas (for
whether there are hotspots of crime that emerge, and only emerge on certain d⁣
week). A step that we have already taken in this direction is the creation of crin⁣
animations to visualise space and time interaction. Each *frame* produce⁣
animation is created following a similar procedure to that described above, wher
frame's parameters and incremental mean thematic thresholds are used as the c⁣
all other subsequent frames (see `http://www.agi.org.uk/cdsig/hotspot`
`time-animation.htm`).

3.7 CONCLUSION

The procedure we suggest creates statistically robust hotspot maps of crime. Pr
analysis tests for clustering and dispersion in our crime point distributic
continuous surface smoothing method (quartic kernel density estimation) w⁣
creates understandable grid cell value outputs that relate to crime, uses paramet⁣
to the spatial distribution of the points being analysed, and derives crim
estimates based on calculations performed at all locations (Levine, 2002), whilst
some practical flexibility in map design. An incremental mean approach helps ⁣
consistently define crime hotspot legend thresholds, producing mapping ou
describes relative levels of crime concentration. From the results on data for th
Borough of Hackney, we suggest a crime density value above 3 multiples o⁣
cells' mean to be a useful defining threshold that describes when a concentratio
hotspot status. This definition of crime concentration can then be supported and
with an advanced option that applies the Gi* local indicator of spatial associatio⁣

Any form of map creation and design requires flexibility. As mentioned, the procedure we suggest retains that flexibility but introduces some simple to apply operations. By including these operations into continuous surface crime hotspot map creation could avoid mistakes in their interpretation and add confidence to the final output.

3.8 REFERENCES

Anselin, L., 1992, *SpaceStat: A Program for the Statistical Analysis of Spatial Data*. (Santa Barbara: National Center for Geographic Information and Analysis, University of California).

Anselin, L., 1995, Local indicators of spatial association – LISA. *Geographical Analysis*. **27**(2), pp. 93-115.

Bailey, T.C. and Gatrell, A.C., 1995, *Interactive Spatial Data Analysis*. (Harlow: Longman).

Chakravorty, S., 1995, Identifying crime clusters: the spatial principles. *Middle States Geographer,* **28**, pp. 53-58.

Eck, J., Drapkin, D. and Gersh, J., 1998, A multi-method exploration of crime hot-spots: An assessment of Vertical Mapper for examining crime concentration. *Crime Mapping Research Centre Intramural Project*, (Washington D.C.: The National Institute of Justice).

Ebdon, D., 1985, *Statistics in Geography*. (Oxford: Blackwell).

Getis, A. and Ord, J.K., 1996, Local spatial statistics: An overview. In *Spatial Analysis: Modelling in a GIS Environment*, edited by Longley, P. and Batty, M. (Cambridge: GeoInformation International), pp. 261-277.

Home Office, 2001, *Focus Areas (Hotspots) Toolkit*. Crime Reduction Unit. http://www.crimereduction.gov.uk/toolkits/fa00.htm.

Illinois Criminal Justice Information Authority, 1996, *Spatial and Temporal Analysis of Crime*. State of Illinois.

Jefferis, E., 1999, A multi-method exploration of crime hot-spots: a summary of findings. *Crime Mapping Research Centre Intramural Project*, (Washington D.C.: The National Institute of Justice).

Levine, N., 2002, *CrimeStat II: A Spatial Statistics Program for the Analysis of Crime Incident Locations* (version 2.0). Ned Levine & Associates, Houston, Texas, and the National Institute of Justice, Washington, DC., April. http://www.icpsr.umich.edu/NACJD/crimestat.html.

Northwood Geoscience, 1998, *Vertical Mapper Version 2.0 Manual*. (Ontario: Nepean).

Openshaw, S., 1984, *The Modifiable Areal Unit Problem. Concepts and Techniques in Modern Geography*, **38**, (Norwich: GeoBooks).

Ratcliffe, J., 1999a, *Spatial Pattern Analysis Machine Version 1.2 Users Guide*. http://www.bigfoot.com/~jerry.ratcliffe.

Ratcliffe, J., 1999b, *Hotspot Detective for MapInfo Helpfile Version 1.0*. http://www.bigfoot.com/~jerry.ratcliffe.

Ratcliffe, J. and McCullagh, M.J., 1998, Hotbeds of crime and the search for spatial accuracy. Paper presented to the *2nd Crime Mapping Research Center Conference: Mapping Out Crime*, Arlington, Virginia, USA. December 10 – 12, 1998. http://www.bigfoot.com/~jerry.ratcliffe.

Rayment, M.R., 1995, Spatial and temporal crime analysis techniques. *Crime Analysis Unit paper*, Vancouver Police Department.

Reid, S., 1999, *Crime: When is a hotspot a hotspot?* MSc dissertation for
    Science Degree in Geographical Information Systems. University of E
    Scotland.

U.S. Institute of Justice, 1999, *CrimeStat: A Spatial Statistics Program for the A
    Crime Incident Records.* `http://www.ojp.usdoj.gov/cmrc/tools/welc`
    `#crimestat.`

Williamson, D., McLafferty, S., Goldsmith, V., Mallenkopf, J. and McGuire, P
    better method to smooth crime incident data. *ESRI ArcUser Magazine,*
    March, pp. 1-5. `http://www.esri.com/news/arcuser/0199/crimedata.ht`

4

Evaluating situational crime prevention: the Merseyside 'alleygating' schemes

Chris Young, Alex Hirschfield, Kate Bowers and Shane Johnson

4.1 INTRODUCTION

It is encouraging to note that examples of mapping and GIS research in the field of crime analysis are becoming increasingly widespread and extensive (see Hirschfield and Bowers, 2001), but to date there has been little reference to the use of GIS for the monitoring and evaluation of *Alleygating* – an intervention designed to both reduce and prevent crime, reduce fear of crime, improve the urban environment, (re)build communities, and improve health. Such schemes on Merseyside attempt to be truly holistic because the gates are also co-designed by the police and constructed by recovering drugs offenders.

This chapter focuses specifically upon the crime reduction aspect of Alleygating, and the use of GIS and police recorded crime data to measure the impact of this intervention on Merseyside – a metropolitan region in the NW of England, (principle city Liverpool situated on the River Mersey). The evaluation is set against the backdrop of a generally downward trend of crime in Britain – the main exception being robbery and theft from the person as reported in the 2000 British Crime Survey – BCS (Kershaw *et al.*, 2000). According to the BCS, burglary was down 21% from 1997 to its lowest level since 1991, the number of vehicles stolen was down by 11% from its 1997 level and theft from vehicles was down by 16%. However the BCS, where possible, also compares the results from their survey with recorded offences and from 1997-99, burglary was down by 13%, and theft from vehicles was down by 9%. Possible explanations for these differences include precision of the BCS estimates and trends in reporting and recording of crimes to the police. However, the overall downward trend still provides a benchmark against which the Merseyside Alleygating evaluation can be compared.

4.2 BACKGROUND TO THE EVALUATION

A 'Northern Evaluation Consortium' from the Universities of Liverpool, Hull and Huddersfield, lead by the Environmental Criminology Research Unit (ECRU) in the Department of Civic Design at Liverpool, was commissioned by the Home Office in

September 1999 to evaluate 21 burglary reduction projects situated in the England. These 21 projects were part of the first phase of 63 country-wide Development Projects (SDPs) of the Burglary Reduction Initiative, an ar government's wider £250 million national Crime Reduction Programme.

The 63 SDPs were funded, after an application process, on the basis that attempting to tackle domestic burglary in ways that adopted a mixture of inte that were either new, or at least applied existing techniques in a novel man instance, in combination with other interventions. The evaluation is still on final outcome and process reports will not be submitted to the Home Office unt

Across such a large number of projects, it was perhaps inevitable that so 21 SDPs would end up trying to reduce domestic burglary by adopting interver were not unique. It was therefore an exciting prospect for the evaluation conson able to compare the application of an intervention (or at least an intervention various 'packages' of interventions) in at least 2 or more different geograp demographic areas. However, an important aspect and first step of the evalua start with individual SDPs. Final results of the 21 SDPs by which their perform be assessed, will be the sum total of each individual project, which in turn is cor several individual interventions. A single intervention type is the subject of the Some example interventions include target hardening, offender supervision surveillance, diversion schemes and market disruption, but it is 'Alleygating offers a particularly attractive option for study here.

Alleygating has been chosen because it is a novel domestic burglary method involving the installation of secure gates at the entrances/exits to All the rear of domestic, usually older Edwardian terraced, properties. An examp such gate is shown in Figure 4.1. All residents (as well as principle utility ser emergency services) are provided with a registered key to the gate that provides their own Alley. An Alley can be of various sizes in order to allow either pede vehicles to gain easy access and exit to and from the rear of properties for both and unfortunately, illegal purposes.

Figure 4.1 Alleygate in-situ.

Data indicates that a high percentage of domestic burglaries are perpetrated via the rear of properties; the British Crime Survey reports 55%, but research specific to mostly terraced streets finds that the figure increases to 72% for this type of housing (Johnson & Loxley – 2001). This is likely to be because of the more concealed aspect of the Alleyway compared to the front of a property in urban areas, where the offender is at greater risk of being disturbed and/or observed. Alleyways also provide a convenient and again, relatively hidden, network by which an escape can be made after an offence has been committed. In addition Alleys are often poorly lit (although 'Alleygate' schemes often attempt to improve lighting at the same time as the gates are installed) and create the perfect environment for other illicit activities such as prostitution and disorder offences including 'youths causing annoyance'. They also become an area for fly-tipping and rubbish dumping allowing domestic pets to rummage through bin bags, leading to obvious unwanted environmental issues.

The mechanics and practicalities of Alleygating from initial plans through to eventual installation is a long and complex process, involving procedures such as highway closure, public rights of way issues and resident consultation and consent, that in itself merits the attention of a dedicated study. Unpublished results from a 'pilot' Alleygate scheme covering some 200 households and four streets in an area of terraced housing in Liverpool on Merseyside indicated that domestic burglary had decreased by around 50% after the installation of the Alley-gates. The pilot area demonstrated other positive outcomes such as reductions in fear of crime, environmental improvements and the formation of an active residents association. This chapter extends the spatial analysis of recorded domestic burglary patterns within Alley-gated zones because it is not limited to just one single scheme. It benefits from a before and after gate installation period in 20 gated areas (including the pilot area) in one region of the UK. By concentrating upon the range of data and GIS techniques required to undertake a robust evaluation of the Alleygating intervention the validity of any results extolling the virtues of Alleygating as a crime reduction technique are also given added weight.

The use of GIS is particularly appropriate in the evaluation of Alleygating because this intervention protects individual properties within geographically defined Alley-gated zones. The area of Merseyside is also appropriate because it is a 'pioneering' Alleygate region within the UK and many of the agencies concerned with process management, and the raw data required for evaluation were already familiar to ECRU. The original 'pilot' scheme was lead by the Safer Merseyside Partnership (SMP) and Merseyside Police but because of the spectacular expansion and popularity of Alleygating throughout Merseyside, an agency (Safer Terraces Alleygating Project – STAP) was established to manage the entire Alleygating process, but with continued input and support from the SMP, Police and other relevant players such as Local Authorities, Fire Service and residents groups.

In summary, this chapter will describe aspects of the evaluation of one particular type of crime reduction intervention in one geographical region; Alleygating in Merseyside. The techniques can be used by the Northern Evaluation Consortium in some of the other SDPs that are attempting to install gates, but importantly they form one component within a toolbox of techniques that can be applied in the evaluation of crime reduction programmes.

4.3   METHODOLOGY

STAP provided ECRU with an Ordnance Survey map for each of the 20 curre
areas. Each map displayed the location of every gate within each Alleygated are
position of these gates in relation to the network of Alleys and streets allowec
defined 'Alleygated' zone to be identified. Each zone was digitised to create a
map containing 20 polygons – each polygon therefore representing an area
from burglary by Alleygates. Instead of having just one Alleygated area the a
now free to choose combinations of 2 or more areas up to a maximum of 20
analysis. This fact is potentially useful because each scheme had a different star
not all were complete, that is, some areas were only partly 'protected' because a
of writing they were waiting for remaining gates to be installed. Hov
completeness and in order to maximise the impact of the data, all 20 gated a
analysed as one 'block'. A larger number of schemes increases the validity of
analysis and although just 6 areas had been fully gated as of 1st April 200(
unreasonable to assume that work in progress in the remaining 14 areas will hav
effects upon the frequency of domestic burglary. Appropriately the police recor
data provided to the author was for financial year periods (*i.e.* April 1st to N
inclusive). Also because, at this stage of the research, the authors are concerne
medium to long-term effects of Alleygating, and not the month-by-month or c
effects, the precise date tag attached to each burglary record within the police
ignored. Therefore GIS analysis within this chapter will investigate higher]
'patterns' in the Alleygated areas. The 'before' Alleygating period is straig
since crime data is available from 1995/6, and the first gates were not installed $
1998. The next phase of installation was between January and October 2000, w
gated at 1st April 2000 as stated above. The crime data for the period April 1999
2000 and April 2000 to March 2001 inclusive would therefore represent
'transition' period rather than an 'after' (in the true sense of the word) A
period. That is, this study had 20 areas at its disposal that were either fully o
gated as of the 31st March 2001. One issue regarding the treatment of all 20 are
single block of protected properties was that standardised buffer zones could [
overlap, but as will be described below, crime rates could be calculated for any
shape of buffer zone since the number of properties in each buffer zone would
based upon property counts within each buffer zone.

4.3.1 Crime Counts

In order to calculate crime counts and rates for each of the Alleygated areas, dat
to domestic burglary and properties were obtained from Merseyside Police and
Survey (via the local planning authority) respectively.

The Merseyside Police Integrated Criminal Justice System (ICJS) has b
since the early 1990s. However the algorithms used to generate grid referen
domestic address information, incorporated within ICJS, were generated durin
1980s and therefore pre-date national standard data sets such as Ordnance
Address Point. Nevertheless, the algorithms developed were quite pioneering
worth dwelling on the process of data capture here, to understand how a grid
for a recorded domestic burglary is obtained. When a crime occurs a police offi
the address into ICJS and the algorithm calls upon a database called MARS (M

Address Referencing System). MARS is a geo-referenced street and address system made up of line and point segments. A long straight road can be made of several segments and curved sections can be made up of many more. Each segment has a known length and a street or road name attribute. MARS also contains a 'per segment' addressed property count. To allocate the crime to a one-metre grid reference along the segment the algorithm first matches the address and then divides the known length of the segment by the known number of properties. Thus a 100 metre line segment with 20 properties numbered 1 to 20 will place house number 5, 25metres along from the start point of the segment. It is clear that this may be suitable for a row of identical terraced properties but is not entirely appropriate for a line segment representing a road consisting of different sized properties such as a mix of semi-detached/detached and terraced houses; the grid reference allocated could fall just outside the boundary of the property it is supposed to represent.

Visual inspections by the authors revealed that most of the burglary crime record grid references were at least allocated to the correct Alleygated zone, if not always the correct property. A problem might arise if one was to research, using GIS tools, crime reduction interventions at the individual property level (for example target hardening) because an error of only a few metres could place an incident in the wrong property. If this was the case, address cleaning processes that deal with data quality issues are available and are discussed by Johnson *et al.* (1997). Cleaned addresses can be matched with a database such as Ordnance Survey's Address Point; a procedure which would then include the allocation of a grid reference from the database, that falls within the property boundary, close to the property centroid.

This research could not use aggregate crime data (that is, at the police beat level) because although Alleygating itself is areal, it is protecting new groups of individual properties that do not conform to sets of existing boundaries such as police beats[1], Census Enumeration Districts (Eds), or local authority wards. Best-fit look-up tables are an option – matching police beat crime data to approximate boundaries of Alleygated zones (based on wards or EDs) but this is clearly not ideal. Disaggregate level, individually geo-referenced crime records are therefore the best option for analysis.

ICJS data provided by the Merseyside Police Force was for complete financial years from 1995/6 to 2000/01 inclusive. Thus a substantial 'before', and a reasonable 'after' Alleygating period (given that the first Alleygate scheme went live in April 1998, the rest beginning to come on-stream during early 2000) of domestic burglary data was available for this research. To comply with data protection agreements and victim confidentiality, data security measures (including secure storage and the use of non-networked computers) were in place at all times when dealing with and analysing the individual-level data derived from ICJS.

4.3.2 Property Counts

To calculate a burglary rate for the Alleygated areas it was necessary to produce property counts for the areas protected by Alley-gates. 1991 Census household data has been used in the past because it takes account of Houses in Multiple Occupancy (HMOs). One property could contain several separate flats or apartments and a burglary rate would effectively be increased if calculations were based on properties. For example, 10

---

[1] Police beats define operational boundaries within which police officers patrol.

burglaries recorded in 100 properties during a year equates to a rate of 100 bur
1000 properties. However if each of those 100 properties contained 2 separate
200 flats or households) the rate would reduce to 50 burglaries per 1000 house
the same period. A decision was take not to use 1991 census household cou
because it is becoming out of date given that Census night 2001 has already ta
However, it could be argued that household – and property counts – especi
measured at the regional level, do not change dramatically, even over a 10 year
more relevant consideration is that household counts are only available down to
and it was felt that an approximation to the digitised Alleygated areas, althoug
using GIS, would be less appropriate than using property counts based upon
Survey's Address Point. Address Point data for 1999 for Merseyside was av;
this research from the local planning authority, and the value of the dataset
highlighted when property counts were required for the buffer zones to be u
analysis of displacement. Rates based on properties would at least be consiste
being artificially inflated in reality if it were possible to compare them with r
upon household counts) since they would be applied to all polygons under inv
including both, Alleygated areas and all buffer zones. In essence, a denor
required in order to ascertain changes in the burglary *rate* year-on-year.

The ICJS recorded crime data was imported into Arcview GIS as a *.dbf* fil
National Grid Eastings and Northings of each crime incident providing
coordinates for the conversion to shape files for each of the financial years fr
to 2000/2001.

Ordnance Survey's Address Point tiles[2] in national transfer format (N
imported using a programme called the 'Classic NTF Translator' pro
'ByDesign' for Mapinfo, and supplied by TMS of Harrogate. This programme
Address Point data into a MapInfo table. The table was then translated from Ma
an Arcview shape file for an area large enough both to cover all of the 20 A
zones and a region extending beyond all 20 Alleygated zones to be used as a
buffer zones. There was no scientific reason for choosing Arcview over Ma
researcher simply had more experience of data manipulation and analysis
former, but found that the NTF conversion programme available for May
particularly user-friendly.

4.3.3 Buffer Zone Analysis

A concentric ring buffer zone was chosen for several reasons, but primarily
allows researchers to make full use of individual level crime data. With aggr
representing police beats for example, the researcher would have to look at
police beats surrounding the selected beat for evidence of displacement. Figure
an example of Alleygated areas in relation to police beat topography, a
demonstrates that the crime reduction intervention will not always occupy an e
Concentric buffer zones also provide the researcher with a standard systema
analyse any potential displacement/diffusion of benefit from Alleygated zones
discussion of the benefits of both concentric buffer zones and disaggregate leve
be found in Bowers & Johnson (2001).

---

[2] A tile is simply a supply unit – for example a 1km x 1km digital map sheet.

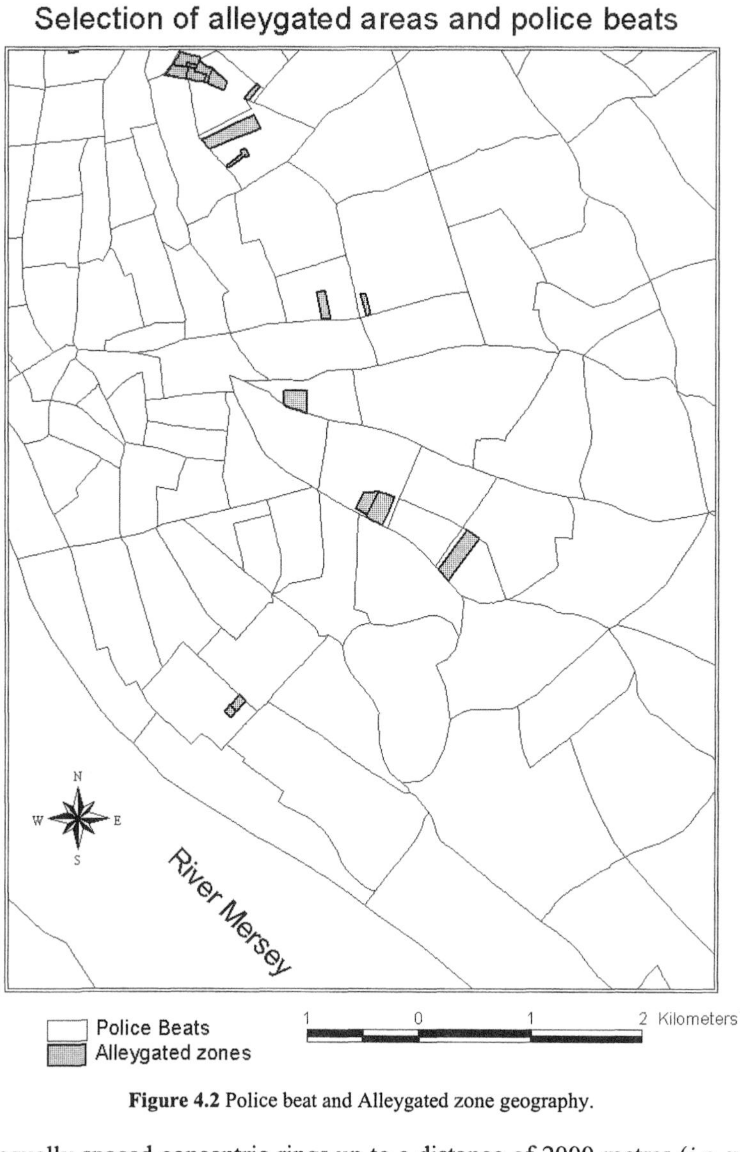

Selection of alleygated areas and police beats

Figure 4.2 Police beat and Alleygated zone geography.

Ten equally spaced concentric rings up to a distance of 2000 metres (*i.e.* under one and a half imperial miles) were chosen because of examples (Wiles and Costello, 2000) in the literature indicating that most crimes of domestic burglary are committed within one mile of the offender's home address. Ten individual rings of 200 metres were chosen so that the area of each ring approximated the areal sizes of the protected Alleygated zones (although this is problematic as they range from just 17 up to 397 properties, but with an average of 173 properties). Figure 4.3 represents these buffer zones geographically and the extent to which they cover a large proportion of the Liverpool district, including the River Mersey which would not, in reality of course, suffer from displacement of domestic burglary! Principal transport routes are also shown to give a representation of scale.

20 alleygated areas and 10x200metre buffer zones

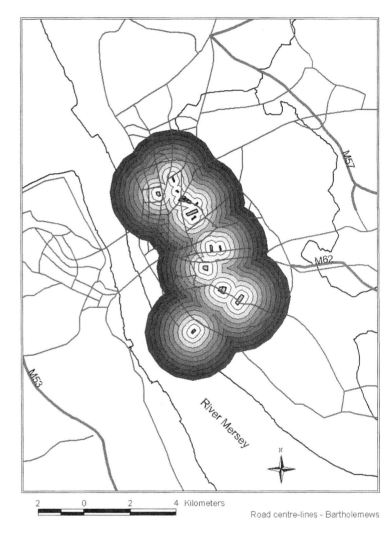

Figure 4.3 Concentric ring buffer zones around Alleygated areas in Merseyside.

By using concentric ring buffer zones it would also be interesting to s
single ring suffered more displacement than any other ring. Indeed it migl
'diffusion of benefits', instead of displacement, into areas surrounding Alleyga
takes place as was the case for a scheme analysed by Johnson *et al.* (2001). In
displacement was found to be worst in the 600 metre buffer zone, with dit
benefit occurring in the 400 metre buffer zone. Street and building topography
buffer zones will probably affect diffusion/displacement – for example domestie
is unlikely to displace into a public park that does not contain any households. L
nearby affluent area containing many large detached houses with adequate

measures might not be the preferred new target of a career criminal more used to breaking and entering via the rear of a poorly defended terraced property in a less affluent area. A diffusion of benefit can occur if domestic burglars perceive that a wider area has been Alleygated than is actually the case. On the other hand even if they know exactly where all of the gates are they may think (mistakenly or not) that other measures have been introduced in conjunction with the gates such as more police patrols for instance.

The buffer zones cover quite a large region of Merseyside but data available for the entire Police Force Area (PFA) of Merseyside (*i.e.* the rest of Merseyside excluding the Alleygated areas and the buffer zones) could be used for comparison (or control) purposes. Data for the PFA would therefore help provide some understanding of the general pattern of domestic burglary in the whole region, within which the Alleygated areas and buffer zones are operating, throughout the six-year period for which burglary data was available.

4.4　RESULTS

Table 4.1 shows the raw counts and crime rates for all of the burglary data available for the 20 combined Alleygated areas at the time of writing. It is based on 3,442 properties in 20-gated areas protected by a total of 208 individual gates, which were erected between January and October 2000 except for one pilot area, which was gated in April 1998. Data provided by STAP indicated that a lesser total of 3,160 properties were protected. This discrepancy is mainly due to the fact that STAP exclude void properties from the protected property count but a contributory factor might be the fact that Ordnance Surveys' Address Point includes non-domestic properties that may not have been included in the STAP 'domestic' property total. Nevertheless the Address Point total of 3,442 properties was used so that crime rates in Alleygated areas would be consistent with those calculated in the buffer zones. Table 4.1 also contains the crime rates for the PFA which shows trends in domestic burglary in the whole of Merseyside between 1995/6 and 2000/01.

Table 4.1 Results in all 20 Alleygated zones (3,442 properties) – plus PFA rates.

Year	Number of Incidents	Number of Repeats	Alleygated Incident Rate	Police Force Area Incident Rate
1995/96	214	32	62.17	32.08
1996/97	185	32	53.75	26.36
1997/98	150	21	43.58	23.08
1998/99	214	27	62.17	22.03
1999/00	206	25	59.85	21.83
2000/01	100	8	29.05	20.05

Table 4.1 was produced using 12 month slices of data only (individual cr dates were not used). It is worth mentioning here that any seasonal fluctuations data – for example an increase in domestic burglary during the Christmas peric therefore be masked. Also, some of the Alleygated areas did not receive October 2000. Therefore despite work to fit Alleygates being in progress October 2000) this fact was likely to have a detrimental effect on the burglar therefore 'contaminate' the burglary data for the period April 2000 to March 2(was supposed to represent an 'after' Alleygating period. Nevertheless A appears to be having a positive effect on the burglary rate within all of the Alley-gate zones because the incident rate of 29.05 in 2000/01 is substantially the previous lowest rate of 43.58 burglaries in 1997/8. The total number o burglary incidents is also significantly better than the previous best of 21, also Put another way repeats make up 8% of incidents in 2000/01 in Alleygated other years between 1995/6 and 1999/00 this figure varied between 12.1% and

Observing the results in all of the Alleygated zones and the PFA show 4.1, against the results for all ten buffer zones combined – Table 4.2 and Figure – it can be seen that national trends of falling burglary rates (Kershaw *et al.*, echoed in Merseyside. There are year-on-year reductions since 1998/9 in areas and buffer zones, but since 1995/6 in the PFA. In the entire buffer zo incidents account for between 8.4% (1997/8) and 10.9% (1995/6) of burglary taking place during any one year, which compares favourably with the fig quoted above for the Alleygated areas in 2000/01.

Table 4.2 Results in all ten buffer zones (109,654 properties).

Year	Number of Incidents	Number of Repeats	Buffer Zone Incident Rate
1995/96	5186	563	47.29
1996/97	3988	381	36.37
1997/98	3754	316	34.23
1998/99	4302	412	39.23
1999/00	4004	427	36.51
2000/01	3486	315	31.79

Figure 4.4 is a graphical representation of the incident rates per 1000 shown in Tables 4.1 and 4.2 respectively for the Alleygated areas, buffer zones It can be seen that 2000/2001 is the first time that the rate in the newly Alleyg: has fallen below the rate in the surrounding buffer zones. This implies that A does have a positive and significant impact upon burglaries within protected z also evident that during the six year period between 1995/6 and 2000/01 bur; overall have fallen by 37.5% in the PFA, 32.8% in the buffer zones and 53. Alleygated zones.

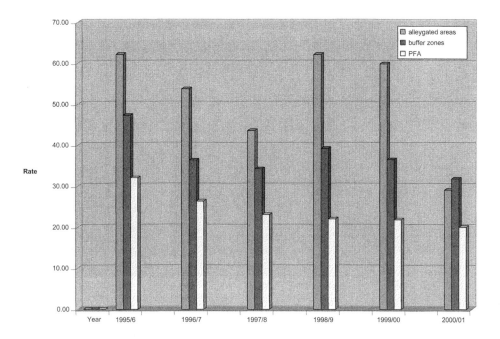

Figure 4.4 Comparison of domestic burglary rate between all Alley-gated areas, buffer zones and the PFA from 1995/6 to 2000/01.

Figure 4.5 shows the difference between the burglary rate in *each* of the 200 metre buffer zones (these figures are not shown but were used to produce the totals in column 2 of Table 4.2) and the burglary rate in the Alleygated areas shown in Table 4.1. The crime rate in the Alleygated zones was subtracted from the crime rate of each individual buffer zone for every year of the six years that burglary data is available.

In Figure 4.5 if a bar is greater than zero it indicates that the rate recorded in the individual buffer zone was higher than the rate recorded in the Alleygated area during that particular year within the period 1995-2001. For the period that preceded the start of the bulk of the schemes (*i.e.* 1995/6 to 1999/00) this situation arises only on three occasions - in the 200m buffer zone in 1995/6 and the 200m and 1000m buffer zones in 1997/8. Following the inception of Alleygating in 2000/01 the first five of the 200 metre buffer zones, from zero to 1000 metres, experience a burglary rate which is higher than that inside the Alleygated zones. Obviously any randomly selected area can have a rate greater than another randomly selected area, but the difference here is that an intervention – Alleygating – has been introduced. Therefore, the data for 2000/2001 favours an argument that the correct areas were selected for Alleygating because there has been an impressive reduction in the burglary rate. Unfortunately it appears that some displacement is occurring from the 20 Alleygated areas into the buffer zones.

It is also apparent from Figure 4.5 that, in descending order, the four zones suffering the most from displacement effects are the 200, 400, 1000 and 600 metre buffer zones. Johnson *et al.* (2001) found that the 600 metre buffer zone was worse off; but this study examined just one SDP with Alleygating, and a known series of further interventions.

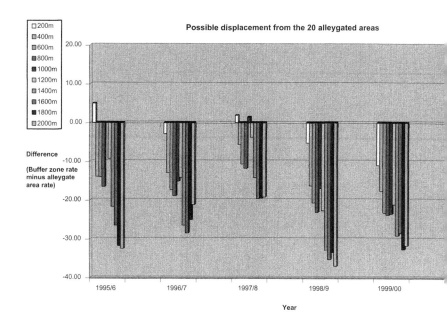

Figure 4.5 Difference between individual buffer zone rates and the Alley-gated area rate for each year between 1995/6 and 2000/01.

4.5 CONCLUSIONS

There are perhaps three main points to be drawn from this research:

- Alleygating appears to be effective in reducing the recorded burglary rate than 50% compared to years when gates were not installed. This reductic to exceed the current national decline of 13% since 1997 in this crime (Kershaw *et al.* 2000).

- Displacement from newly Alleygated areas is an issue that requires at practitioners, especially if surrounding areas consist of similar housing demographies.

- GIS can be an effective tool for the planning and management of Al schemes. This chapter has demonstrated the usefulness of GIS for the n and evaluation of Alley-gate interventions particularly because th reduction/prevention initiative, despite being clearly defined and geog concentrated, does not conform to any pre-existing boundary or land parce

The burglary rates for the Alleygated areas shown in Table 4.1 vary betw and 62.17 – a reduction of almost 54% during the period 1995/6 to 2000/01. previous 'best' of 43.58 burglaries per 1000 properties in 1997/8 has been imp some 33%. At the same time the domestic burglary rate in the buffer zones (1 ranges between 31.79 and 47.29 – a reduction of almost 33% during the period

2000/01. However the previous 'best' of 34.23 burglaries per 1000 properties in 1997/8 has only been improved by 7.1%.

Displacement is an issue that cannot be ignored but other studies are optimistic that diffusion of benefits can also result from the installation of Alley-gates (Johnson *et al.*, 2001).

Using GIS gives the researcher freedom to explore different avenues in an evaluation, but does require the availability of detailed and accurate data. The only drawback might be that results published outside of the police service have to be re-aggregated for data protection and confidentiality purposes, but at least the GIS means that a researcher can choose the size and type of area and is not restricted to statutory boundaries.

4.6 FURTHER RESEARCH

It would also be interesting from both a research and a social policy effectiveness point of view, to analyse a further 12 or even 24 months of data to March 2003 for example, to ascertain the true sustainability of Alleygating as a crime reduction intervention. However consideration will have to be given to the fact that more Alleygated areas will become 'live' during that time and some may be sited inside the buffer zones of Figure 4.3, therefore affecting the results. Repeat burglary figures were shown in Tables 4.1 and 4.2 and although they are also encouraging in terms of a reduction, they merit a more in-depth analysis in their own right.

A fuller evaluation, now that more and more areas are being Alleygated should include a more detailed look at the crime data. The modus operandi (MO) of domestic burglary in both new and established Alleygated areas requires investigation. Do burglars simply switch to another mode of entry other than the rear of the property or two or three years down the line do residents simply become complacent and leave the gates unlocked allowing burglars to gain easy access again?

An analysis of the temporal pattern of domestic burglary (*i.e.* year-on-year, month-to-month, weekly or even daily) in Alleygated areas might also prove to be a useful exercise. For example, criminals might simply begin to take advantage of the periods when gates are left open to allow refuse collectors to gain access.

Displacement to other types of acquisitive crime (crime switch) should also be considered. With access to domestic properties restricted will offenders attempt to carry out more street robberies or thefts of/from motor vehicles for instance, or will the figures for attempted burglary/criminal damage simply go up if it's harder to get into a house? Burglars may even up the stakes by being prepared to use violence against a resident during a burglary, or use confrontational techniques such as distraction burglary – therefore aggravated burglary or assault figures might increase? Alleygating may well have to be implemented at the same time as complementary vehicle crime reduction campaigns for example, in order to be truly successful. Conversely researchers may have to try and isolate Alleygating as a crime reduction method when trying to evaluate its specific successes or failures. In this chapter no account has been taken of any other crime reduction campaign that may have been operating concurrently (either in the Alleygated areas or the buffer zones) – perhaps a target hardening scheme or intensive offender supervision programme had a larger impact upon the decrease in domestic burglary?

To support the work discussed above, a cost effectiveness study of Alleyg as a process and crime reduction tool would be appropriate but this should not human benefits – what price for improved peace of mind and maybe even be for the residents of these areas? Resident surveys, health impact questionnaire from other agencies such as the Fire Service should help complete the bigger the Alleygating phenomenon.

4.7 ACKNOWLEDGEMENTS

This research is based upon the initiatives of the Safer Merseyside Partnersh authors are grateful for their co-operation and support. Photograph of courtesy of Safer Terraces Alleygating Project.

4.8 REFERENCES

Bowers, K.J. and Johnson, S.D., 2001, Measuring the geographical dis of Crime. *Submitted to the Journal of Quantitative Criminology*.

Hirschfield, A. and Bowers, K. (editors), 2001, *Mapping and Analysing Cr Lessons from Research and Practice*. (London: Taylor and Francis).

Johnson, S.D., Bowers, K., and Hirschfield, A., 1997, New insights spatial and temporal distribution of repeat victimisation. *The British J Criminology*, **37**(2), pp. 224-244.

Johnson, S.D., Bowers, K.J., Young, C.A. and Hirschfield, A.F.G., 2001, Unco true picture: Evaluating crime reduction initiatives using disaggregate cr *Crime Prevention and Community Safety: An International Journal,* in press

Johnson, S.D., and Loxley, C., 2001, *Installing Alley-Gates: Practical Les. Burglary Prevention Projects*. Home Office Policing & Reducing Cr Briefing Note 2/01. (http://www.homeoffice.gov.uk/rds/prgpdfs/brf20

Kershaw, C. *et al.*, 2000, *The 2000 British Crime Survey*. Home Office Development and Statistics Directorate.

Wiles, P. and Costello, A., 2000, *The Road to Nowhere: The Evic Travelling Criminals*. Home Office Research Study No 207. London: and Reducing Crime Unit, Home Office Research, Development and Directorate.

5

Crime hot spot prediction: a framework for progress

Jonathan Corcoran and Andrew Ware

5.1 INTRODUCTION

Crime rates differ between types of urban district, and these disparities are best explained by the variation in use of urban sites by differing populations. A database of violent incidents is rich in spatial information and studies have to date, provided a statistical analysis of the variables within this data. However, a much richer survey can be undertaken by linking this database with other spatial databases, such as the Census of Population, weather and police databases. Coupling Geographical Information Systems (GIS) with Artificial Neural Networks (ANNs) offers a means of uncovering hidden relationships and trends within these disparate databases. Therefore, this paper outlines the first stage in the development of such a system, designed to facilitate the prediction of crime hot spots. For this stage, a series of Kohonen Self-Organising Maps (KSOMs) will be used to cluster the data in a way that should allow common features to be extracted.

5.2 CRIME PREDICTION

The advent of computers and the availability of desktop mapping software have advanced the analytical process, allowing efficient generation of virtual pin maps depicting crime incidents. A logical step beyond visualisation and analysis of trends and patterns is the development of a predictive system capable of forecasting changes to existing hot spots and the evolution of new ones.

Crime prediction in criminological research has been established for a number of decades (Ohlin and Duncan, 1949; Glueck, 1960; Francis, 1971), although its foundation within a geographic and GIS context is still in its infancy.

As predictive crime analysis is a new research area, very little literature currently exists. Olligschlaeger (1997) provides an overview of existing forecasting techniques, concluding the time, level of user interaction and the expertise that each demands would be unrealistic for implementation in an operational policing environment (Olligschlaeger and Gorr, 1997). In addition, the inherent inflexibility and inability to dynamically adapt to change would compromise their viability in policing. ANNs are presented as one technique that offers minimal user interaction in addition to dynamic adaptability, and thus a potential operational forecasting solution.

5.2.1 Potential for Crime Prediction by the Police

A recent survey (Corcoran and Ware, 2001) has highlighted the uptake a
computer based crime-mapping technologies by British Police Forces. Con
mapping technologies are rapidly becoming a vital prerequisite for visuali
incident distributions and assisting in both the identification/allocation of reso
production/evaluation of policing strategies. The ability to efficiently genera
maps to depict crime location and densities can be used directly to inform polic
and policing strategies, therefore maximising effectiveness and potential. A rec
published by the Home Office (Home Office, 2000) has underlined the impo
geographic data for analysis and interpretation of crime at the local level. In ad
Chief Constable of Kent County Constabulary notes that *"over the last few yea
activity has shifted its centre of balance away from the reactive investigation aft
towards targeting active criminals on the balance of intelligence"* (Phillips, 200
A natural step beyond visualisation and analysis of past and current incide
prediction of future occurrences, thus providing an *early warning system* for t
(Olligschlaeger and Gorr, 1997). Prediction can help prevent crime in that it
the optimal allocation of limited resources. Prevention might be better than cu
the real world, very often, this is under financial constraints. The developmen
for prediction can thus help prevent crime and maintain optimal operational cost
Prediction requirements for the police have been classified into th
categories according to the period of time involved (Gorr, Olligschlaeger *et al., 2*

- Short-term (tactical deployment);
- Medium-term (resource allocation);
- Long-term (strategic planning).

The focus for crime forecasting lies with short-term models as police
respond to existing and emerging crime patterns on relatively short time-s
example on the basis of daily, weekly and monthly figures (Gorr and Ollig
1998). This paper details a potential prediction framework for short-term
deployment of police resources. The framework will allow identification of ri
from which probabilities of criminal activity (specifically emergence of hot spo
derived and the necessary resources deployed.

5.2.2 COPDAT

The Crime and Offender Pattern Detection Analysis Technique (COPDAT), o
this paper, offers a potential framework that can be applied to geographic p
COPDAT entails the implementation of a GIS, integrating various spatial dat
analyse and map the identified trends.

5.3   METHODOLOGY

The volume of crime is insufficient to accurately predict an offence (in terms o
and time) when extrapolating from past offences. Therefore, in the proposed sy
type of crime predicted to take place within a particular time-window is supplen
a separate prediction of the likely vulnerable areas for the same epoch. In ac

would seem prudent to have the facility in a finished system to allow information based on police intelligence and experience to be built into the predictive model. The idea is to enhance current police predictive capabilities not replace them.

The spatial framework for the prototype COPDAT conforms to police sector boundaries for Cardiff (the capital city of Wales in the United Kingdom), whereby the volume of crime is sufficient to derive accurate predictions.

5.3.1 Data Sets

The accurate forecasting of the temporal-geography of crime (where and when a crime is likely to take place) would be of immense benefit, for accurate prediction if acted upon should lead to effective prevention. However, crime prediction frequently relies on the use of data appertaining to past perpetrators and/or past victims. Such data is therefore subject to legal and ethical restriction on its use, resulting in an ethical conundrum (Ware and Corcoran 2001). Therefore, the prototype COPDAT involves the use of only two central data sets – one pertaining to crime and the other providing contextual information.

5.3.2 GIS Techniques

Visual inspection of the various spatio-temporal data sets is a vital pre-requisite in assimilating an understanding of fundamental relationships and trends. The GIS is used as a tool to conduct this basic pattern analysis, including cluster and hot spot techniques.

5.3.3 Artificial Neural Network Techniques

ANN models provide a mechanism through which the various spatial, non-spatial and temporal data sets can be analysed to identify patterns and trends previously undiscovered. Identification and consolidation of such trends and patterns between the various data sets allows generalisation and subsequent prediction of future events and scenarios based upon that generality. For example, identifying that a specific kind of crime is likely to occur in a certain location type under a given set of ambient conditions allows future incidents to be predicted.

The result of this scenario can be to produce a spatial probability matrix encapsulating a risk assessment of the study area. The spatial probability matrix can be directed to the GIS for visualisation in the form of a thematic contour map differentiating different areas with respect to their potential risk. In essence, the ultimate goal of the COPDAT is to learn decision criteria for assigning risk levels to new and future situations. This, for example, may involve identifying and predicting areas most at risk during a hot summer bank holiday football match in the City. Provision of such information is of obvious interest and of operational benefit to the police in terms of both resource allocation and policing strategies.

5.3.4 Data Preparation and Processing

Data representation and structuring is of key importance in the production of a robust predictive model. It has been shown in previous neural network studies that a certain

level of pre-processing of the raw data is advantageous to model accuracy, effic
stability. This approach subsequently requires a certain level of post-processin
to generate the required output values (illustrated in Figure 5.1).

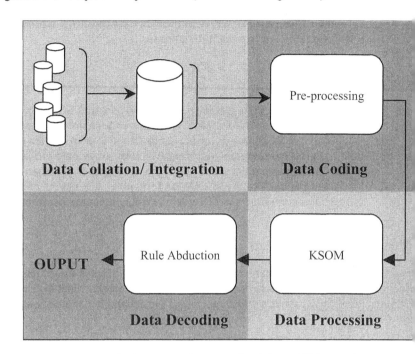

Figure 5.1 Data preparation process in relation to ANN processing.

5.3.5 Data Pre-Processing

First, the data will undergo systematic and comprehensive analysis. This is
where necessary, by converting the data into an alternative representation more
input into an ANN. This sequential process can be broken down to a series o
stages:

- Format conversion and integration;
- Error detection and editing;
- Data reduction, transformation and generalisation.

The final stage in the pre-processing is of critical consequence in te
successful ANN implementation. The process of feature extraction and encodin
characteristics impacts upon the ANN's ability to learn and assimilate rela
between salient variables and hence its accuracy in prediction. This proces
further decomposed into three distinct procedures:

1.   Transformation and scaling may include:
     - Applying a mathematical function (*e.g.* logarithm or square) to an input
     - Scaling the different inputs so that they are all within a fixed range ca
       effect the reliability of an ANN system.

2.   Reduction of relevant data includes simple operations such as filtering combinations of inputs to optimise the information content. This is particularly important when the data is noisy or contains irrelevant and potentially erroneous information.

3.   Encoding of identified features for input to the ANN. The data types include a range of data categories (discrete, continuous, categorical and symbolic), each to be handled and represented in an explicit manner.

5.3.6 Clustering Using a Kohonen Self Organising Map

Temporal, spatial and incident data will be clustered using a series of KSOMs. These clusters, whose data share the same characteristics, will form the basis for rule abduction. (Note, deduction is where the effect is deduced from the cause – for example, 'the burglary was committed because the house was left unlocked.' Abduction is where the cause is gleaned from analysing the effect – for example, 'sun and alcohol often causes socially unacceptable behaviour'.)

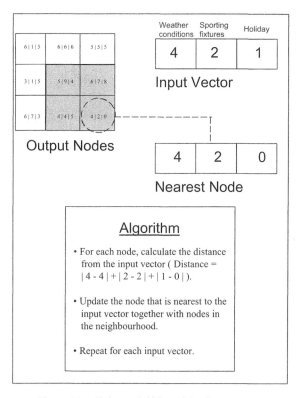

Figure 5.2 A Kohonen Self Organising Feature Map.

An unsupervised network, such as the Kohonen Self Organising Map (KSOM), organises itself in such a way as to represent classes within a data set. The 2D KSOM allows classes to be visualised on a feature map, in which similar inputs are spatially clustered. Figure 5.2 shows a typical 2D KSOM along with an abridged algorithm (N.B., the number of nodes are arbitrarily selected for example purposes).

Each output node on the KSOM contains a vector of length j, where j is eq number of input attributes. Before training begins, the network is placed initialised state, *i.e.* the directions of the vectors in each node are randomised. involves passing an input vector into the network through the input nodes. Each the KSOM is then compared with the input vector, and the closest node is chan more like the input vector. Neighbouring nodes also become more like the inp Iterating this process achieves clustering of similar input vectors in Euclidean sp.

5.3.7 An Overview of the Methodology

The methodology uses a KSOM to find clusters in the input vectors and then from each cluster is used to train a separate MLP network. The advantage of KSOM for this application is that it can identify clusters within the parent datase difficult to identify using simple sort procedures. Figure 5.3 gives an overvie method. A dataset containing the required elements of the vector is passed thr KSOM during the training stage and allowed to develop into clusters. After tra clusters are inspected and the primary clustered features used to describe the sub These sub-datasets are then used as the basis for rule abduction.

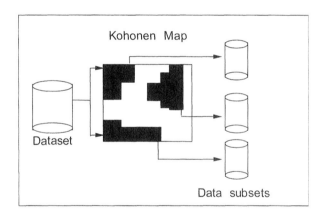

Figure 5.3 Using a KSOM to partition a dataset.

However, two fundamental problems need to be resolved before this methc of any use. First, the problem of separating adjacent clusters, and second, the proceed to the abduction phase only using 'good' clusters (see Figure 5.6).

The first problem has been recognised in other studies and some guideli been provided (James 1994). In essence, the problem lies in the attribution of nodes to a specific cluster. Figure 5.4 provides an example of a KSOM ou adjacent clusters. There appear to be four classes within the dataset, but there ar of uncertainty relating to the boundaries of each cluster (the Digits in each node number of vectors mapped to that node).

To overcome this problem a simple method of identifying class boun discriminants can be used, which relies on the fact that a KSOM clusters prir binary features. For example, if the type of crime is represented using binary in KSOM will tend to cluster records according to this attribute. Boundaries adjacent clusters on a 2D map can then be found by inspecting the records m each node and grouping together nodes that contain the same classification values. However, this level of clustering can be achieved using a multi-level sort procedure. In essence, the binary representation of the data will dictate the make-up of the resulting clusters and more importantly the homogeneity of the data sets.

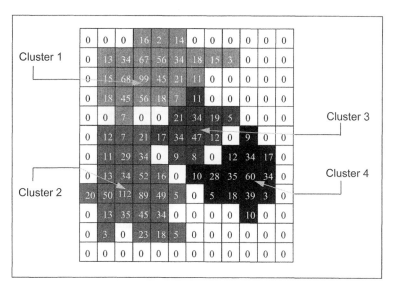

Figure 5.4 An Example Trained Kohonen Self Organising Feature Map.

If the data are represented using continuous inputs, the clusters formed by the KSOM would provide more generalised classes which would be difficult to achieve using a sort procedure. However, the inspection method would no longer identify class boundaries, as the similarities between records would not always be apparent. Clearly, before meaningful training data sets can be formed, the problem of discerning effective class boundaries in a KSOM must be addressed. Ideally, the network adaption rule should cluster similar inputs and clearly distance individual clusters. Zurada (1992) *explains "One possible network adaption rule is: A pattern added to the cluster has to be closer to the centre of the cluster than to the centre of any other cluster".* Using this rule, each node can be examined and the distance from the surrounding centroids can be calculated (a centroid is taken to be a node, outside any known cluster boundaries, that has the largest number of input vectors mapped to it). The subject node can then be added to the nearest cluster. Figure 5.5 illustrates a hypothetical situation where it is unclear where to draw the boundaries around clusters on a KSOM.

By simply calculating the Euclidean distance of the subject node from the two centroids, the subject node can be assigned to the closest cluster. However, in this application, which aims to generate clusters with latent but meaningful information that can be subsequently extracted using abduction techniques, the formation of a class boundary for Cluster 1 (including the subject node) may dramatically increase the variance of the training data. This increase will reduce the potential accuracy of the model. In the example, it may have been better to exclude the subject node from either of the clusters, and deem the vectors mapped to the subject node as either being outliers or a separate cluster.

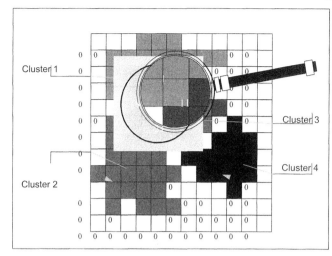

Figure 5.5 An Example KSOM.

In addition to identifying boundaries around input clusters, it is also imp
this application to match input clusters to appropriate output clusters. In terms o
activity, if, for example, the KSOM has clustered crimes from two different
areas, it is reasonable to expect these crimes to have similar characteristics.

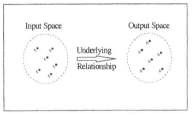

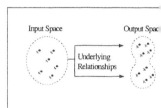

(a) An Example of a Good Input Cluster. A one-to-
one relationship can be established and hence Input
Space is homogeneous.

(b) An Example of a Bad Input Cluster.
more similar vectors in the Input Spac
different vectors in the Output Space. H
Input Space is not homogeneous.

Figure 5.6 Example cluster mappings from input to output space.

Figure 5.6a illustrates a cluster of similar input vectors. When the corre
data in output space is examined, all the examples describe similar outpu
Conversely, Figure 5.6b shows a situation where the data can only be modelled
or more functions. The problem now is to estimate the 'usefulness' of a give
There are a number of options available of which the following are the most use

- Multi-Layered Perceptron (MLP) Model (Chen 1997)
- Class Entropy (Quinlan 1986)
- R^2 Almy (Almy 1998)
- Gamma Test (Stefánsson 1997)

For classification problems, Class Entropy can be used to decide if input clusters are homogenous with respect to output clusters. For example, Quinlan's C4.5 and C5.0 (Quinlan 1993) uses Class Entropy to segment the input space until each segment points toward a single class in output space. However, this approach is not applicable for regression problems such as this one and this rules out the use of class entropy.

A quick and easy estimate of the susceptibility of a dataset for function induction can be achieved by executing a multiple regression analysis on the data and use the R^2 value to discern trainable clusters. This technique is useful for data where the function is known to be linear. However, this is not known to be true for crime analysis data.

The **Gamma Test** attempts to estimate the best mean square error that can be achieved by any smooth modelling technique using the data. If y is the output of a function then the Gamma test estimates the variance of the part of y that cannot be accounted for by a smooth (differentiable) functional transformation. Thus if $y = f(x) + r$, where the function f is unknown and r is statistical noise, the Gamma test estimates Var(r). Var(r) provides a lower bound for the mean squared error of the output y, beyond which additional training is of no significant use. Therefore, knowing Var(r) for a data set allows prediction beforehand of what the MSE of the best possible neural network trained on that data would be. The Gamma test provides a method of determining the quality of the data stratification – a good stratification technique will result in a low value of Var(r) for each subset. Interpreting the output from the Gamma test requires considerable care and attention.

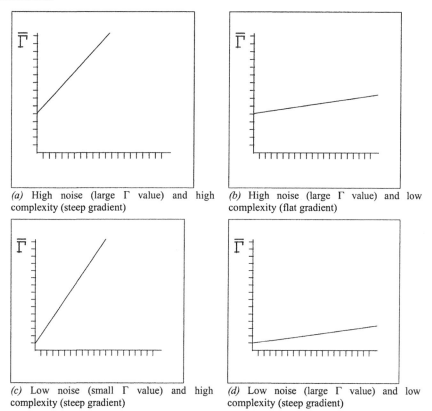

(a) High noise (large Γ value) and high complexity (steep gradient)

(b) High noise (large Γ value) and low complexity (flat gradient)

(c) Low noise (small Γ value) and high complexity (steep gradient)

(d) Low noise (large Γ value) and low complexity (steep gradient)

Figure 5.7 Interpreting the output from the Gamma Test.

The least squares regression line provides two pieces of information. intercept on the Gamma axis is an estimate of the best MSE achievable by an modelling technique. Second, the gradient gives an indication of the complex underlying smooth function running through the data. The Gamma test may e very low MSE but unfortunately show a high level of complexity that could be to accurately model. It is easier to see this situation when the output from the test is presented graphically.

A hypothetical example with high noise content and high complexity is Figure 5.7a; high noise and low complexity Figure 5.7b; low noise and high co in Figure 5.7c; and, finally, low noise and low complexity (the desired ou Figure 5.7d. In summary, for this methodology to be successful, the foll required:

- class boundaries must be identified around clusters formed by the KSOM input space that exclude outliers and nodes from neighbouring clusters, and

- only 'good' clusters (illustrated in Figure 5.6) should go on to form trai sets for subsequent back propagation models.

5.3.8 A Detailed Look at the Methodology

The Gamma test can be used at a number of abstraction levels within the stratification method:

- Cluster level;
- Node Level;
- Record Level.

Data stratification is achieved at cluster level or at node level, dependir ease at which cluster boundaries can be determined. The record level gives an i of outliers.

5.3.8.1 Cluster Level Analysis

This can be achieved thus:

Identify Cluster boundaries in KSOM
For every cluster
  Place records mapped to cluster into a file
  Apply Gamma test to data in the file
  If Var(r) <= some Threshold then
      Use data file as the training set for a MLP
  else
      Process at Node Level
  Endif
Next Cluster

This level of abstraction is the least computationally intensive as it only one pass of the Gamma test for each cluster. The disadvantage with this method is often difficult to identify boundaries manually. In this case the Gamma test s applied at the Node level.

5.3.8.2 Node Level Analysis

At this abstraction level, the methodology attempts to identify useful clusters by selecting a centroid and adding neighbouring nodes – where the addition of a node increases the variance significantly it is subsequently removed. This process iterates until the cluster size is maximised within a specified variance threshold. This algorithm identifies useful clusters on a 2D KSOM. This is achieved thus:

```
number_of_clusters:=0
While there are nodes to cluster
    number_of_clusters: = number_of_clusters + 1
    Select the unclustered node with the largest record count
    Apply Gamma test to the data in the selected node
    If Var(r) <= Threshold then
        nodes_of_interest:= None    (cluster includes only the data from selected node)
        For each unclustered node next to selected node
            Add data from unclustered node to the cluster
            Run gamma test on cluster
            If Var(r) <= Threshold then
                Add node number to nodes_of_interest
            else
                Remove data from the unclustered node from the cluster
            Endif
        NextNode
        While nodes_of_interest <> None
            Select c_node from nodes_of_interest
            Remove c_node from nodes_of_interest
            For each unclustered node immediately surrounding c_node
                Add data from unclustered node to the cluster
                Run gamma test on cluster
                If Var(r) <= Threshold then
                    Add node  to nodes_of_interest
                else
                    Remove data from the node from cluster
                    Record the boundaries of this cluster
                Endif
            NextNode
        EndWhile
    else
        Process at Record Level
    Endif
EndWhile
```

The boundary detection algorithm for a 1D KSOM is very similar except neighbouring nodes are selected progressively further away from the left and the right of the centroid node. This level of analysis is more computationally intensive than the cluster level analysis, as it requires $m*\sum n_i$ passes of the Gamma test, where i is the number of nodes investigated for cluster n for a KSOM containing m clusters. If using either the cluster level analysis or the node level analysis, useful clusters have been identified, it is then possible to train an independent MLP on each subset. The KSOM is then used to select the appropriate MLP on which to predict the value of a previously unseen example. The resulting system is closely related to a panel judgement system. However, if both methods have still resulted in poor training sets (useless clusters) then the analysis is taken to the most detailed abstraction level, that is the record level.

5.3.8.3 Record Level Analysis

The record level analysis is the most computationally intensive. The purpose of of the methodology is to identify data subsets from examples that have map same node on the KSOM. This facilitates extraction of outliers from a data set giving some indication as to the examples that require additional features. The developed for this level of analysis is very similar to that shown for the n analysis. However now, sets of records are iteratively analysed using the Ga This is achieved thus:

> *For each node in the KSOM*
> > *Apply Gamma test to estimate the variance for the data in node*
> > *If Var(r) > Threshold then*
> > > *For each record at node*
> > > > *Remove record from data set*
> > > > *Apply Gamma test to estimate the variance for the data in node*
> > > > *If New Var(r) < Previous Var(r) then*
> > > > > *Mark record as outlier*
> > > >
> > > > *else*
> > > > > *Add record back into data set*
> > > >
> > > > *Endif*
> > >
> > > *NextRecord*
> >
> > *else*
> > > *Proceed at Node Level*
> >
> > *Endif*
>
> *NextNode*

This level of analysis will identify the need for additional features and records that may be classed as outliers.

5.3.9 Post-Processing

Natural language rules will be derived from each 'good' cluster found by the Each rule will represent a broad and generic definition (that with time can be fi of a specific sub-model that can be applied to best predict crime, for example:

> *if Centre of City*
> *and Weather includes Wet*
> *and Day is Friday*
> *and Time is Night*
> *then Problems will include inside pubs (probability 0.9)*

> *if Centre of City*
> *and Weather includes Dry*
> *and Day is Friday*
> *and Time is Night*
> *then Problems will include inside pubs (probability 0.4)*

> *if Centre of City*
> *and Weather includes Dry*
> *and Weather includes Warm*
> *and Day is Friday*
> *and Time is Night*
> *then Problems areas will include outside pubs (probability 0.9)*

These rules together with other determining factors (including temporal information such as the day, time and prevailing weather) are then directed to the GIS for production of a thematic crime risk contour map. When used in a predictive manner, the GIS can provide an important visual reference for analysing the relative impact of multiple factors on crime levels in a given area.

5.4 IMPLEMENTATION DETAILS

A series of KSOMs will provide the mechanism for classifying the heterogeneous nature of crime and criminal activity in a spatial, temporal and contextual framework. The assumption that there is a single simple linear relationship between the various factors is unrealistic and misplaced. In the light of this, a valid approach to assimilate an understanding of crime and criminal activity is to dissemble it into its homogeneous components for independent analysis. This analysis can be used to generate a series of rules and generalisations that can be utilised in a predictive capacity at a broader heterogeneous level.

The KSOM provides an unsupervised technique, through which this stratification process can be automated, resulting in the formulation of homogeneous clusters (illustrated in Figure 5.8). Where clusters are not clearly delineated it maybe necessary to calculate a function such as the Euclidean distance or Gamma test (Lewis, Ware *et al.* 1997b, 1997a) in order to assign output to specific clusters.

The user will be able to present the trained suite of KSOMs, in the COPDAT, with structured scenarios from which probabilities and vulnerabilities of criminal activity will be derived and presented to the user via spatial thematic representations. Using standard GIS functionality, the user could subsequently overlay additional information, which may for example involve identification of specific localities, (street names, building etc) for deployment of resources.

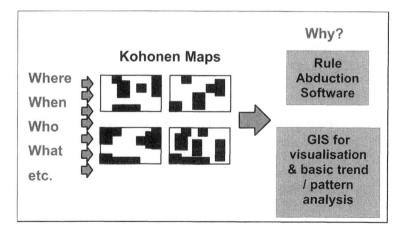

Figure 5.8 Proposed COPDAT Hybrid System.

5.5 SUMMARY AND BENEFITS

Coupling the predictive capabilities provided by the proposed COPDAT m
existing GIS techniques would enable law enforcement agencies to more e
evaluate resource and tactical policies. Crime mapping and analysis using COP
potentially facilitate research, assist with offender management and monito
enable community planners to develop policy and forecast future needs for pub
resources. Provision of such facilities will rely upon use of multiple data sourc
dependent upon the use of pre-defined data processing and encoding technique:
require new standards and techniques to be integrated into current practices. 1
Constable of Kent County Constabulary raises this point in respect to police in
noting that *"this process is only possible if we can mix and match information e*
board. To this end, it is essential that common standards and discipline atta
intelligence process" (Phillips 2000).

   This paper outlines a potential framework that can be expanded to m
predict a broader variety of crimes based upon a larger mixture of criminal and c
data sets and a common schema for crime prediction.

5.6 REFERENCES

Almy, R., Horbas J., Cusack, M., Gloudemans, R., (editors), 1998, *The Vai
    Residential Property Using Regression Analysis.* Computer Assisted Mass *
    An International Review, (England: Ashgate Publishing Company).
Chen, K., Xiang,Y., Huisheng, C., 1997, Combining linear discriminant funct
    neural networks for supervised learning. *Journal of Neural Compu
    Applications* **6**.
Corcoran, J. and Ware, J.A., 2001, *Application of Computer Crime
    Technologies by British Police Forces.* Available from: jcorcora@glam.ac.t
Francis, S.H., 1971, *Prediction Methods in Criminology*, Home Office Research
Glueck, S.E., 1960, *Predicting Delinquency and Crime.* (Cambridge, Mass:
    Harvard University Press).
Gorr, W. and Olligschlager, A., 1998, *Crime Hot Spot Forecasting: Mode
    Comparative Evaluation.* http://www.ojp.usdoj.gov/cmrc/grants/welcon
Gorr, W., Olligschlaeger, A. and Thompson, Y., 2000, Assessment of crime fc
    accuracy for deployment of police, *International Journal of Forecasting.*
Home Office, 2000, *Review of Crime Statistics: A Discussion Document.*
    www.homeoffice.gov.uk/crimprev/crimstco.htm.
James, H., 1994, *An 'Automatic Pilot' for Surveyors*, RICS Cutting Edge.
Lewis, O.M., Ware, J.A. and Jenkins, D.H., 1997a, A novel neural network tech
    modelling data containing multiple functions. *Computational Intelligence
    and Applications*, Springer Verlag, pp. 141-149.
Lewis, O.M., Ware, J.A. and Jenkins, D.H., 1997b, A novel neural network tech
    the valuation of residential properties, *Journal of Neural Computing and Ap
    5, pp. 224-229.
Ohlin, L.E. and Duncan, O.D., 1949, The efficiency of prediction in cri
    American Journal of Sociology: LIV 5.

Olligschlaeger, A. and Gorr, W., 1997, *Spatio-Temporal Forecasting of Crime: Application of Classical and Neural Network Methods.* http://www.heinz.cmu.edu/wpapers/active/wp00162.html.

Phillips, D., 2000, *National Intelligence Model*, National Criminal Intelligence Service.

Quinlan, J.R., 1986, Induction of decision trees, *Machine Learning* **1**, pp. 81-106.

Quinlan, J.R., 1993, *C4.5: Programs for Machine Learning*, Morgan Kaufmann.

Stefánsson, A., Koncar, N., Jones, A.J., 1997, A note of the Gamma test, *Neural Computing and Applications* **5**, pp. 131-133.

Ware, J.A. and Corcoran, J., 2001, *Forecasting Crime: An Ethical Conundrum.* Available from: jaware@glam.ac.uk.

Zurada, J. M., 1992, *Introduction to Artificial Neural Network Systems*, West Publishing Company.

PART II

GIS and Planning

6

E-community participation: communicating spatial planning and design using web-based maps

Kheir Al-Kodmany

6.1 INTRODUCTION

This chapter addresses innovative methods in using Web-based mapping and Geographic Information Systems (GIS) in public participation and community planning. The ultimate goal is to explore how Web mapping can be advanced to facilitate two-way communication between planners and the public. Presently, most Web-based mapping systems use one-way communication to provide static information to the public. Current GIS technologies lend themselves well to one-way communication. While this type of system can be very useful, it is important to consider new technologies that offer opportunities to use the Web in ways that have not previously been recognised. Using examples from a university-community partnership between the University of Illinois at Chicago (UIC) and community groups in Chicago's Pilsen and North Lawndale neighbourhoods, several such new technologies are introduced and examined. These examples show the progression from one-way communication to two-way and even three-way communication using Web-based maps. The challenges of creating interactive screen-based maps for public participation include navigating large geographic area maps on a small screen, selecting specific map features, geo-referencing public input and transferring this input to conventional GIS systems. These challenges as well as many successes of our Web mapping research are explored in detail. As Web-based mapping capabilities are refined and improved, new avenues are opened for public participation in the planning process.

The research in this chapter reflects recent views about the importance of incorporating citizen participation into computer-based planning efforts. The phrase 'public participation GIS', or PPGIS, describes recent research from the planning profession that is rooted in the concern that all voices should be heard in a democracy. In particular, it aims at improving access to GIS among non-governmental organizations and individuals, especially those who have been historically under-represented in public policy making. Individuals and citizens' groups without access to GIS may find it

difficult to gain entry into a public policy-making process that relies on GIS
difficult to challenge policies that were created through the use of 'expert' GIS
Advocates of PPGIS claim that GIS technology does not adequately repres
societal groups. Some researchers and practitioners instead seek to develop a
GIS systems (called GIS/2) that are more adaptable to input from citizens
official sources (Obermeyer, 1998b). In this model, the role of participants in
and evaluating data is primary. GIS/2 systems seek to accommodate an
representation of diverse views (Aitken and Michel, 1995; Rundstrom, 199
1995; Obermeyer and Pinto, 1994; Obermeyer, 1995; Al-Kodmany, 2000).

This chapter also follows the framework first developed by Lynch (196
(1998), and Sanoff (1991). These scholars advocate users' need-driven de
planning or a user-oriented approach. They argue that current practices do not a
address users' needs and preferences. The research described here is propell
belief that public needs are underrepresented in today's planning and design pra
that Web technology offers a medium of communication that can potentially ad
problem. The goal is to determine a strategy that allows residents to voice their
existing conditions or on proposed plans and development schemes and to sh
views amongst themselves and with community organizations and authorities.
has become common for local governments, community planners and ce
agencies to utilize the Web to offer information to community members, th
takes the next step by using the Web as a medium for two-way communication
allows people to become both receivers and providers of information. This pr
out some of the potential contributions and limitations related to enabling
communication and participation over the Web.

The University of Illinois at Chicago, where this project originated, aims a
mutually beneficial relationships among and between faculty, scholars,
community leaders, institutions, and researchers. It promotes engaged urban
that relies on collaboration and partnership between the university and indivi
organizations in the public and private sectors. The communication among th
groups is a model for the type of two-way communication that we strive for v
based technology in planning. Two-way interaction is critical for suppc
university-community partnership. In two-way communication any partner may
and/or receive communication, and as a result, a cycle of communication, or
knowledge is generated. Two-way communication enables each partner to bec
a consumer and a producer of knowledge and opens the possibility for sha
knowledge with neighbouring communities throughout Chicago, the region
world.

The project described in this chapter consists of a number of prototypical
that represent a variety of attempts to expand upon present Web-based
technology to a new kind of interactive mapping. Each application we have dev
intended to provide a structured process to gather information from multiple pa
and then to present these views pictorially. The goal is to help the Unive
communities think more effectively as a group focused on a common endeavo
losing their individuality, and to help Community Based Organizations (CBO)
the complexity of their ideas without trivializing them or losing detail.

In developing the one-way, two-way and three-way spatial communica
address several challenges in creating Web-based participatory maps. One chal
been to create a way to access user spatial feedback without using a GIS syste
GIS software such as ArcIMS, demands long download times and substantial ba

and its interface design is not intuitive, in our view, we have developed interactive websites using other technologies.

Other challenges involve efforts to replicate traditional, paper-based participation methods on the computer. For instance, we have experimented with various ways of representing and navigating a large geographic area on a small computer screen. Unlike paper maps, computer screens limit the amount of geographic area that can be viewed at one time. Scrolling across a large map online is usually impossible because of the slow speed and large amount of computer memory that is required to reload the images. We have developed two interfaces that present alternative ways of presenting a map of a large community area without excess complication or computer resource use.

Another challenge is developing a flexible, accurate way for users to select and comment on a specific neighbourhood, block, or building. In one Web interface design, we used pre-defined square units on a map (using a grid system) so that it would be possible to easily analyse the location selection data from a large group of participants. However, we subsequently found that this method could excessively limit user choices and so we developed a new interface that features freeform drawing as the method of selection. The development of Web-based freehand drawing tools, where users can use an online 'pen' to select their own locations, is a particularly exciting innovation. In both of these research areas we have experimented with the use of aerial photographs, structure base maps and GIS maps of the community as the background upon which users can select locations of concern. After presenting several options for navigation and selection, we discuss one interface where we attempted to incorporate the best features into one 'complete' or integrated interface. It will be most helpful to read this chapter in front of a computer in order to try out the interfaces. It will be most helpful to read this chapter in front of a computer in order to try out the interfaces. The Internet Explorer browser is recommended for best results.

6.2   THEORETICAL BACKGROUND:
IMAGEABILITY AND PUBLIC PARTICIPATION IN PLANNING

The theoretical foundations for these Web interface designs can be found in the work of Lynch (1960) and Nasar (1998), who argue for the importance of discovering how city design affects citizens. By studying what kinds of evaluative images residents have of their community, planners, researchers and community leaders can derive valuable information about how to improve the physical form of their communities.

In his seminal book, *The Image of the City* (1960), Lynch uses the concept of 'imageability' as a theoretical framework for studying cognitive maps, urban form, and the spatial relationships of cities. Imageability "is that quality in a physical object which gives it a high probability of evoking a strong image in any given observer. It is that shape, colour, or arrangement that facilitates the making of vividly identified, powerfully structured, highly useful mental images of the environment. It might also be called *legibility*" (Lynch, 1960, 9). If most people like the imageable elements of a city, then it will probably convey a positive evaluative image. If they dislike them, the city will convey a negative evaluative image, suggesting a need for changes in the city's appearance. This aspect of city image is what Nasar calls the *likability* of the cityscape. Likability refers to the probability that an environment will evoke a positive evaluative response among the groups of people experiencing it. Inhabitants of a city with a good evaluative image find pleasure in the appearance of its memorable and visible parts.

The city landscape may be a source of pleasure and delight to its resi
visitors, and can potentially counteract the stresses of daily life. Thus, the dev
of a city's form should be guided by a visual plan that is concerned with visua
the urban scale (Lynch, 1960). However, to devise such a plan, we need to knov
public evaluates the cityscape and what meanings they find in it.

Nasar suggests that it is possible to learn the public's preferences by e
measuring them. Just as we weigh objects to find how light or heavy they are, I'
that we can measure preferences to determine the degree to which people like
various areas of a city. Nasar developed and implemented a method of
residents (using traditional phone surveys and manual map-making) to determ
areas they like and dislike in their community, with the goal of creating
'evaluative image' of the community that could guide future design and develop

In the Web-based mapping projects described in this chapter, we have ad
advanced Nasar's original method by using the Web and GIS to survey and ma
preferences. Most of the interfaces we have developed are based on the c
surveying a group of community residents to discover their preferences and
about the imageability of their community. GIS is in the background, in the se
functions as the method for analysing the data that is collected via the Web. TI
used as the primary platform for collecting spatial information.

6.3   CURRENT WORK ON PUBLIC PARTICIPATION AND THE INTE

The Internet has already proven to be valuable on its own as a low-cost
communication for participatory planning though Web sites, email, surveys an
conferencing (Craig 1998, Al-Kodmany, 2000). The Neighbourhood Knowl
Angeles project (http://www.nkla.sppsr.ucla.edu) provides a strong examp
the Internet is being used as a communication tool for empowering the pub
project, a collaboration of the municipality of Los Angeles and UCLA, knits t
municipal databases and inspection records, looks for indicators of urban decay,
the information on city maps posted on its Web site. The project utilizes off
software for database management (Microsoft Access and Internet Informatio
and for Internet mapping (ESRI's MapObjects and Internet Map Server). Their
provide public access to government records and electronic mapping throu
computers and at 'touch-screen' information kiosks.

The Internet is now able to support interactive programs in a manner
stand-alone GIS and stand-alone hypermedia systems. Peng (1999) conclude
speed of technological development provides an opportunity to expand GIS te
and spatial information to the general public. While the technology is not quite
"it is likely that we will see an increasing number of distributed GIS with m
components which are organized around a spatial data infrastructure and
through wide-area networks such as the WWW" (Shiffer 1998 p. 731). Kry
that the WWW has great potential for implementing Public Participation Vis
(PPV) and Public Participation GIS (PPGIS) (1999). Recent developments in W
programming languages are making highly interactive advanced GIS ap
available to anyone with a modem and Internet browser. Even novices c
geographic information, amend and add information, and interactively explore
scenarios.

The East St. Louis Action Research Project (ESLARP) is one of many examples of community organizations utilizing this new technology. ESLARP (http://www.imlab. uiuc.edu/eslarp/) is a reciprocal learning effort between members of the East St. Louis community and students and faculty at the University of Illinois. Their primary goal is to help community-based development organizations increase their planning, design and development capabilities while educating planning and design students. One result of the partnership is EGRETS, a geographic information retrieval system designed to be used by community residents. This allows the public to interact with GIS data through a web browser without ever owning GIS software on their machines. In EGRETS, residents can either search for maps already made or create their own maps.

The Internet as a medium of communication will be increasingly utilized in all aspects of planning. It is valuable on its own as a low-cost mode of communication for participatory planning but it becomes particularly powerful when it is used to distribute and disseminate other visualization technologies. While there is great excitement about future possibilities for Internet-based public participation, concerns generally centre around access to the technology. First, access must be ensured in terms of making sure the pool of participants has Internet access so that there will be wide representation in public participation forums, and second, in terms of creating a critical mass of users of a particular Website to sustain meaningful interactions.

Shiffer and others at MIT have researched a variety of ways of using the Web for urban planning and design. They have explored how emerging information technologies and Web technologies can improve the processing and communication of planning-related information in metropolitan planning organisations. They provide case studies illustrating how to deliver spatially-referenced multimedia material for site planning and reviews using projects in Washington DC, the South Boston Seaport and Boston's waterfront development. Such multimedia interfaces, coupled with the accessibility of the Web, have the possibility of opening up a new paradigm within urban design, one which helps to communicate ideas and developments to other agencies and the general public (Shiffer, 1995).

As Shiffer's work has demonstrated, one result of this move toward digital visualisation of urban form and distribution of information on the Web is that there are new possibilities for involving the lay public in design decision-making. As the number of households with Web access increases, and the demographic profile of Web users diversifies, the potential for using the Web for public participation planning increases exponentially. Thus, the exploration of which types of digital tools and interfaces can best engage the public in planning activities is a promising avenue for research.

A survey is one important tool for public participation on the Web. Citizens can use a Web-based survey to become information creators, rather than passive recipients of information. This is an important leap forward in using the Web as a communication tool. Most of these applications utilise simple feedback forms where users type in comments in response to questions and then click 'submit' to send their responses to the Web server. There are fewer examples of Web-based surveys that utilise graphics, maps or other kinds of visualisation to inquire about the public's locational preferences. One example is the Landscape Scenic Preference survey developed by the Macaulay Land Use Research Institute, which aims to quantify the landscape preferences of the general public. Participants are shown pairs of landscape images and asked to choose which they prefer (http://www.mluri.sari.ac.uk:80/~mi550/landscape/). The same survey was conducted in a traditional manner where people were shown photocopied photographs and then asked to give each landscape an evaluative score. Early indications have shown

that the results from the paper-based questionnaire were not significantly diffe
the on-line version. The researchers identified areas that needed improveme
initial reaction was that the on-line survey worked well.

In another example, Kingston (1998) has developed several projects th
Web to facilitate public input on several environmental problems i
(http://www.ccg.leeds.ac.uk/democracy/). In one project, a Web-based
making environment was developed which allows the public to model a n
possible planting scenarios in locating areas for regeneration of native
(Kingston, 1998).

The drawbacks of using the Web for communication between planners,
and the public centre around the broad issue of access. First, though access to tl
increasing, it is still difficult to draw a random, representative sample since Wel
not yet accurately reflect the real demographic makeup of the general public. A
must be quite motivated to log-on and find the Web site in order to participate.
underlies a concern about creating a critical mass of users to sustain meanin
interactions. This prompts the question of what factors are needed to achieve
level of activity (Shiffer, 1995). The examples given above suggest that V
surveys need to be further refined but have now become quite feasible as public
tools.

6.4 BACKGROUND TO OUR PROJECTS

6.4.1 The University of Illinois Chicago

In 1993 the University of Illinois at Chicago (UIC) established the Great Cities
(http://www.uic.edu/cuppa/gci/about/index.htm) to respond to urban
facing American cities. 'Great Cities' refers to the university's commitment t
teaching, research, and service programs to improve the quality of life in met
Chicago. Under the Great Cities program, UIC has worked with mainly
communities on approximately 220 different projects and programs.

One of the newest urban outreach initiatives under Great Cities is the Gr
Urban Data Visualisation program (GCUDV), established in 1998. GCUDV's r
to explore how advanced visualisation capabilities can be used in community pl
create innovative computer applications and to test them in actual urban
(http://www.uic.edu/cuppa/udv/). Researchers in the GCUDV program wo
databases, GIS, and 3D visualisation into a medium suited for urban and
planning and policy. Projects focus on social and ecological data as well as urb
visualisation. The program is staffed by a multidisciplinary team that include
students and recent graduates from UIC's Electronic Visualisation Laboratory
several other disciplines such as Architecture, Art History, Information Techno
Urban Planning.

6.4.2 The Communities

Our research focuses on two Chicago communities: Pilsen (Lower West Side) a
Lawndale. Pilsen, the first of the three communities we work with, has long se
port of entry for many of Chicago's immigrants. By the end of the 19th centu

industrialisation and urbanisation had transformed the largely Bohemian (Czech and German) working-class neighbourhood into a national centre of labour activism. This activity drew Poles, Croatians, Lithuanians, Italians and other immigrants to the community. After a few decades, these residents moved on, giving way to newer Mexican immigrants. In 1990, 88 per cent of the area population was Hispanic, and Pilsen had the second highest concentration of Mexicans of all community areas in the city. Interestingly, each of the successive groups has, in turn, left its unique imprint on the architecture of the community, creating a cultural mosaic in the built form of Pilsen.

Despite being a welcoming home for new immigrants, Pilsen presently struggles to retain residents; many people start out in the community but eventually move on to other neighbourhoods or the suburbs as they assimilate into American culture and become financially secure. The number of housing units in the community has declined and little new residential construction has taken place since 1930. In 1980, 27 per cent of the housing units were overcrowded, having more than one person per room (CFBC, 1990). Over the years, several strong community organisations have formed to help Pilsen's disenfranchised residents address issues like housing and economic development.

The second community, North Lawndale, is located northwest of the university, four miles from downtown and near the United Centre sports stadium. This community, like Pilsen, experienced successive waves of immigrants, beginning with Bohemians and Polish, and followed by a wave of Russian Jews. However, the population decreased by approximately 30,000 people per decade between 1960 and 1980. The 1990 total population was 47,000, a decline of 23 per cent from 1980's total of 61,000. Ninety-six per cent of its 1990 population was African American (CFBC, 1990).

The number of housing units in North Lawndale continues to decline, causing a housing shortage. The community has experienced a net loss of almost half of its housing units since 1960. North Lawndale has an abundance of redbrick and graystone homes, mostly built in the 1910s and 1920s, and though the community has suffered from urban blight, the housing stock has proved to be surprisingly solid for renovation. Recently, the area has begun to see some revival, as one development corporation has turned the old Sears catalog complex into homes at Homan Square, which brings infrastructure and housing improvements to 55 acres. New businesses and families are moving into Lawndale and the economy is starting to turn around. However, the community still has many vacant lots and abandoned buildings.

6.5 METHODOLOGICAL APPROACH

Most of the Web-based GIS sites that are currently available on the Web provide one-way information delivery. Although providing proper spatial information is an important aspect of participatory planning; however, it lacks engaging the public in the planning process. In order to become a robust tool for use in participatory planning, Web-based GIS should go one step further: it should be a medium for two-way spatially based information exchange. Interactive Web-based GIS could become a critical and widely used tool to gain important feedback in community planning.

The general intent of most of the interfaces described below is to discover how residents of Pilsen evaluate the appearance of their community by asking what particular places they like and dislike. The survey instructions, while not included in all of the interface prototypes, emphasise the issue of *community appearance* and encourage respondents to think of which places they find to be pleasant and which places they find

to be blighted and unattractive. The results are intended to be used as guid
decisions on community appearance and visual form. The map tools h
developed to work on average computers and do not require the loading of g
information systems (GIS) or Internet map-server (IMS) applications. Local
simply interact with Web-maps by clicking map locations and typing text to giv
opinions.

Conceptually, the prototypes discussed here are organised into three c
one-way, two-way, and three-way spatial communication. We present the dev
of the tool from simple Web sites that display spatial information to more con
interactive sites that actually allow participants to draw and comment on map
community. The increasing levels of sophistication can be described as fo
simply displaying a map (such as disseminating the results of a spatial an
viewing a map with different types of data included, and some ability to nav
map, 3) maps that include some GIS functions so that users can retrieve attri
along with the map and 4) map surveys that allow users provide feedback direc
maps (Kim, 2001). The degree of user interaction is associated with the type of
software that are used, such as HTML, plug-ins, Java applets, Java Servlets, an
Map Server technologies.

6.5.1 One-way Spatial Communication

One-way spatial communication describes a paradigm by which information is
to an audience without any possibility for interactive response. There ar
approaches to one-way communication that help to make a large geogra
navigable online, including map view, map overlay, nesting maps with i
resolution and the use of thumbnail photographs.

6.5.1.1 Viewing Maps Online

Viewing maps online is a basic function and the simplest to deliver. It is a
communication method in which information is transferred in only one pre
direction. Examples of maps that we used for the purpose of neighbourhood
include site and neighbourhood plans, survey maps, topographical maps, cen
and tract maps, transportation and land use maps, utilities map, and aerial photog

By using simple HTML or Web authoring tools, we are able to publish th
in a one-way static format. A very minimal level of interaction is provided by
browser, which allows scrolling right-left and up-down to view the different pe
the map. Maps are published directly on the Web page after preparing them in a
compatible image format, such as GIF or JPEG. These maps can be snapsho
from digital map, exported using electronic software or scanned in from a pa
The Web site, http://www.uic.edu/~kheir/community_maps/main.html
examples of GIS maps of two neighbouring communities, Pilsen and North L
These maps include locational, ethnicity, population, median income and me
information for these communities. These maps were produced using U
(Housing Urban Development) Community 2020 Maptitude GIS software. La
were exported as GIF image from the application and embedded and hyperlin
HTML document. The client and server architecture adopted here is exactly sa

basic framework of the client and server model. When the client requests maps embedded in regular HTML documents, the Web server pass them to the browser in the client computer (Figure 6.1).

Figure 6.1 Viewing maps on the Web requires a simple client-server architecture.

6.5.1.2 Map Overlay

Overlay analysis is a classic technique that has been discussed extensively by numerous geographers and planners (McHarg 1963, Hopkins 1975, and Edwards, 1984). It is an excellent method for analysing spatial and component relationships. Various facets of a developing concept can be synthesized using successive layers. Physical and non-physical considerations can be fused and viewed graphically, as this procedure illuminates more complex relationships. The map layers act as transparent acetate sheets that, together, form a composite image. By overlaying individual images, in which each details different pieces of information for the same geographic location, one can better understand complex relationships between different types of spatial data within an area. That level of analysis may only be available by examining the overlaps between the various images. Moreover, computerized image manipulation on the Web can be done in ways that have never been possible by traditional graphic means.

By using Java Applet layers, we simulate the GIS overlay function on the Web. The user can interactively turn layers on and off to visualize spatial information and relationships. With this method, there is no need to use ESRI GIS ArcIMS. The Web enables users to display and interact with thematic layers intuitively. The Web site, http://www.uic.edu/~kheir/layer/p1/, provides examples of using a Java applet for overlay function. Layers include the background of the Lakefront community as well as maps showing census tract, median rent, median value, and median income information.

6.5.1.3 Nested Maps

This Web survey (http://g015.cuppa.uic.edu/gridFeedbak/final/index.html) attempts to evaluate participants' likes and dislikes for the four square-mile community of Pilsen. In earlier versions of this Web map/survey tool, we had the problem of not being able to zoom on the map. We wanted to publish the map at the highest possible resolution, yet this took up too much space on the screen; users had to scroll up and down and right and left to see the map. In this new interface, where we are trying to cover a much larger geographic area, this was a serious problem. We found a solution in the concept of a geographic hierarchy with three zooming levels. This Web interface utilises a grid and an aerial map of the community. Users are able to navigate the map by utilising three zooming levels, each with increasing resolution.

When the user selects one of the 16 squares on the initial map (the first level), the program zooms in, and the selected square appears as a new full-screen aerial map that is again broken down into a grid of 16 squares (the second level). When one of the 16 cells

is selected from the second level, an aerial close-up of the selected area (the t⌐
then loads into the window. The actual geographic size of the final selection i⌐
blocks. The highest resolution is on this third level with six inches per pixel. At
the user has the ability to zoom in and out between the three levels. At the t⌐
users can see the fine details and distinguish buildings and streets. In a later
further developed the above Web site to allow two-way communication by
selection and textbox tool for participants to type in comments about the locatio⌐

6.5.1.4 Thumbnails

Another option for navigating a large community area is to show a number of v⌐
'thumbnail' photos, or tiles, which, upon clicking, bring up a larger map of th⌐
demonstration of this technique can be found at `http://g015.cuppa.`
`feng/project.html` and `http://g015.cuppa.uic.edu/ramki/image.html.`
other interfaces, this one is designed using an aerial map of the Pilsen communi⌐
(miniature photographs) are displayed across the top of the screen with some str⌐
visible to help orient participants. The instructions read, "*Select one of the ⌐*
photos by clicking on it. In the text box below the map, describe any physical c⌐
the community that have occurred since the map photo was taken 10 years ago.
only one possible use of thumbnails; other uses could also be envisioned. A
tiles could be used to cover a greater area of the community (however, this
down the interface).

6.5.1.5 Zooming

The two prototypes described above utilise different methods of 'zooming' o⌐
based map, and each has pros and cons. The first method, using nested maps, ⌐
In the Java interface we write the scripts for zooming. There are four Java
different scales (zoom). Once you click the zooming function on the map yo⌐
layer after the other, simulating a 'zoom' on the photograph. When the user cli⌐
the next map (zoomed version) the new map is called from the remote ser⌐
disadvantage of this method is that it can be extremely slow, particularly for u⌐
modem connections. The advantages are that Java provides great flexibility i⌐
and adding new features.

The second method (the thumbnail interface) uses Flash Shockwave softw⌐
is Web design tool to create animated Web pages. The strength of this tool
handles images in vector format. Thus, users do not lose resolution no matter ho⌐
they zoom in on the map. This is a very fast method of zooming since the f⌐
available within the program itself. Zooming is done via a small movie file th⌐
the zooming easy and in real time. Once a user enters the Web site, a messag⌐
asking the user to install Flash software. If the user clicks yes, it loads in j⌐
seconds. The Web site contains a small movie file that automatically gets load⌐
user's hard drive and then the zooming is done from the user's own machine
very fast (Figure 6.2). While speed is a significant advantage of this method, th⌐
Flash software package does not allow the incorporation of features from other ⌐
Thus, we cannot mesh this site with other functions available with the Java prog⌐
language (Figure 6.3). For instance, at this time there is no way to include a
feature so that users can select a building within the photo.

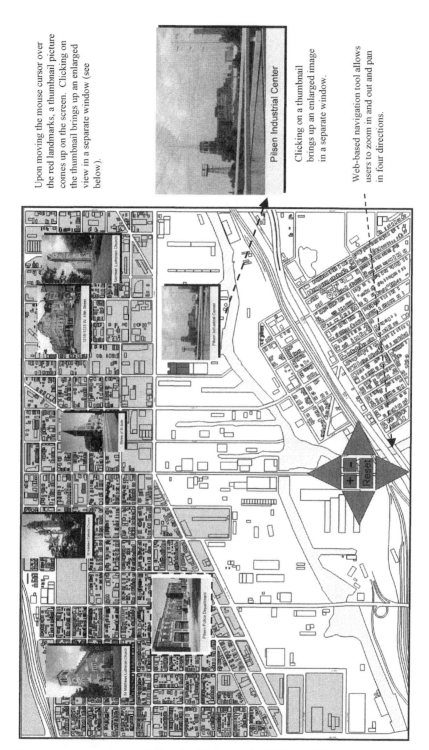

Upon moving the mouse cursor over the red landmarks, a thumbnail picture comes up on the screen. Clicking on the thumbnail brings up an enlarged view in a separate window (see below).

Pilsen Industrial Center

Clicking on a thumbnail brings up an enlarged image in a separate window.

Web-based navigation tool allows users to zoom in and out and pan in four directions.

Figure 6.2 A map constructed using GIS and published on the Web with Macromedia Flash Software to provide navigational capabilities.

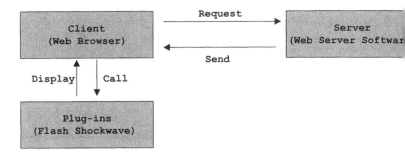

Figure 6.3 The relationship between the server, the client computer, and plug-ins in a zooming en

6.5.1.6 Map and Data Retrieval from GIS

Map and data retrieval refers to enabling users not only to navigate maps b
examine attribute data and perform spatial queries. The purpose is to bring si
functionality on the Web without the need for GIS software at the end user
However, this technology continues to provide limited spatial function on the
user cannot perform advanced analytical functions but can perform simple on
turn on or off layer, set active layer, examine attribute data related to map fe
basic spatial queries. Commercial Internet mapping software from major ven
as ESRI ArcIMS, Intergraph GeoMedia Web Map, AutoDesk MapGuide, and
MapXtreme, are appropriate for performing simple GIS functions online. The
Web site (http://e036.cuppa.uic.edu/arcviewims/north-lawndale/in
shows an example of using ArcView IMS software for delivering spatial infor
the North Lawndale community (Figure 6.4).

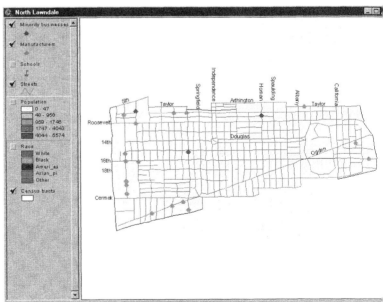

Figure 6.4 ESRI ArcView IMS Information System for North Lawndale community. The view
manufacturers and minority business thematic layers.

Map retrieval enables users to retrieve maps and related attribute data from the server at runtime. The Web server, interpreter, and GIS software work together for this. Whenever users change the view or perform spatial query, the Web server pass it to the GIS software via the interpreter. The GIS software performs the tasks and passes the outcomes, via the interpreter again, to the Web server. Then, Web server sends final outcome in HTML format to the client. A key part in this framework is GIS software. What implements user request spatially is the GIS software. The Web server only receives user requests and sends final outputs. However, since the GIS software cannot directly communicate with the Web server, a middleware is required. The middleware acts as an interpreter for the Web server and GIS software. It passes parameters received by Web server to the GIS software and the results of the task done by the GIS software to the Web server. The Common Gateway Interface (CGI) has often been used for this, but Java Servlet technology is rapidly replacing CGI (Figure 6.5).

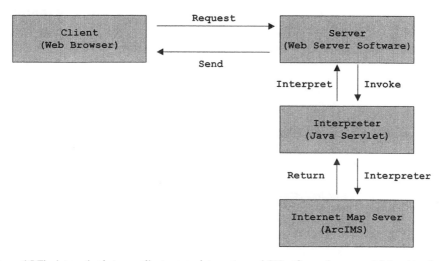

Figure 6.5 The interaction between client, server, interpreter, and GIS software for map and data retrieval.

6.5.2 Two-way Spatial Communication

Getting user feedback on the Web is becoming increasing popular. Sending text input or polling choices is widely used to get users' opinions. However, when working with spatial information, it is important to be able to visualize such feedback. In two-way communication, we want to allow users to annotate maps and to delineate their concerns on maps. One option is to allow users to draw various features, such as points, lines, circles, and polygons on the map. It is also possible to add text along with drawn features. When users send their input to the server, the server detects the x, y coordinates of such drawings and saves them. This makes it possible to define a particular location. Users can indicate exactly which location they are talking about.

Such a spatial feedback system may be designed with or without GIS software. However, at the time of this writing, the only solution that receives spatial user feedback on the map is GIS software, specifically, ArcIMS. The workflow on the server side is the same as in the data retrieval method. However, the function of drawing on the map requires Java applets and Java and ArcIMS plug-ins in the client side. The server and client must work together heavily for this method. The following Web page

(http://e036.cuppa.uic.edu/arcims/north-lawndale) shows an example o
communication using ArcIMS (Figures 6.6 & 6.7).

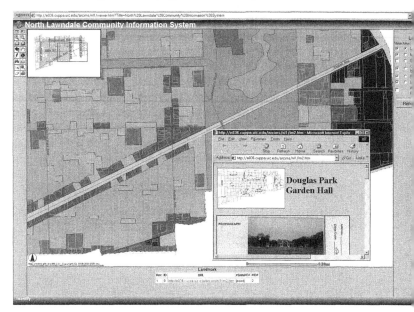

Figure 6.6 ESRI ArcIMS Property Information System for North Lawndale Community. In the
Web-site window that is hyperlinked to the system. It provides additional information abou
properties, such as historic sites in the community.

Figure 6.7 Two-way spatial communication. ESRI ArcIMS Property Information System for Nor
Community. System enables participants to type comments on map and submit them to the Universi

However, ArcIMS has substantial drawbacks, particularly at the client side. It requires a thick client and a heavy server. A 'thin' client means a client computer with just a Web browser, while a 'thick' client implies a computer with a Web browser with other add-ins, such as Java applets, ActiveX controls, and plug-ins to evoke special effects. The thick client needs to download such add-ins at runtime or beforehand. A 'light' server is a Web server computer having only HTML documents and related files, but a 'heavy' server has other components, such as database applications and GIS software working together with the Web server. For ArcIMS, the server requires heavy processing with GIS software and a Java Servlet, and the client also needs a Java plug-in installed beforehand and Java applets at runtime to enable the drawing functions (Figure 6.8).

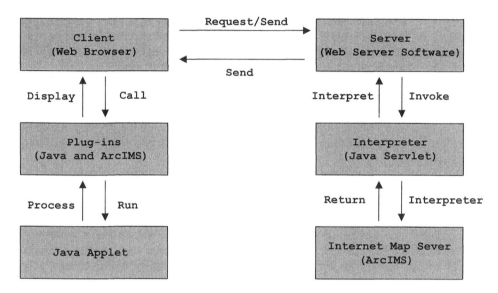

Figure 6.8 Complicated client-server configuration required for ESRI ArcIMS.

In ArcIMS, the download time for data, map and attribute data retrieval is quite long and it requires substantial bandwidth. Most importantly, the interface design and interaction behaviour is not intuitive. Because of these drawbacks of ArcIMS, we have invested research in developing alternative methods for two-way communication. We divide them into four categories: grid-based, freehand, a combination of grid and freehand or a 'complete interface', and a compositional method.

6.5.2.1 Grid Based

We have created several different prototypes that utilise a grid as the underlying selection structure behind a given map. Below we describe three different interfaces that utilise the grid-based selection method.

Our earliest project (http://go15.cuppa.uic.edu/gridfeedbak/xarial18St/arial_18St.html) was a survey consisting of one exercise titled, *Urban Likability and Dislikability* (ULD) for Pilsen. Simply speaking, participants logged on to the project Web site where they could view a high-resolution aerial photograph of Pilsen with a grid overlaid on top of it. We limited the geographic area to the vicinity of the ▮ commercial district, since this area was the primary focus of revitalisatic Participants were asked to point out areas on the map that they liked and dislil provide the reasons for their responses. Each square of the grid was identi. centroid (the centre of the square). This centroid was coded as the actual lon; latitude of the centre of the square.

Participants were asked to identify the areas of their community that they ▮ and disliked by clicking on the appropriate square of the grid. The only visual on the map was the name of streets. They were to use a GREEN pointer to indi▮ areas and a RED pointer to indicate disliked areas. By clicking on one of buttons located on the side of the map, they could 'load' the pointer with eith▮ When participants clicked on the square, a small window with a question mark asking them to state their reasons for liking or disliking that area of the co▮ When finished, the participant clicked on a button labelled 'submit' and their transferred to the UIC server. Each of the participants' selections could then be the Web access logs for analysis and feedback into the planning process.

Since the server was linked to an Oracle database and a GIS application, w capability of taking all the points that were selected by the participants, sortin; longitude and latitude, and plotting them on a map automatically. The Oracle could also group the associated comments. In this manner, a community-input was created that contained the range of views about areas liked and disliked associated reasons.

We then created a number of GIS choroplethic maps to illustrate the in▮ likes and dislikes (urban likability and urban dislikability). We used dots to intensity: the number of dots in each cell of the map was proportionate to the ▮ times that area was selected by the residents in the survey exercise. In addit▮ GIS maps were interactive; clicking on an area (or cell) of the map opened a v text that listed the residents' reasons for liking or disliking that area. Since provided written evaluations for each point, they were extremely useful in specific directions for improvement and could easily be incorporated into the n▮ of the Pilsen community planning process.

In a later prototype (http://g015.cuppa.uic.edu/gridFeedbak/fir .html) (Figure 6.9) we used nested maps. We also used a grid but it advanced project by dealing with a larger geographic scale. Not only does the user navi; the grid, but he or she also selects a square of the grid to comment upon. As ▮ above, the primary disadvantage of this method is that users cannot select th▮ buildings or combination of buildings that they wish to comment upon, since t the final selection square is predefined. The advantages are clearly evident in example where the uniform selection areas allowed us to easily create sort and a users' feedback.

In evaluating the grid as a selection mechanism, it has clear advant▮ disadvantages. As in the above example, the grid enables very fast ana▮ compilation of spatial data that can be easily compared among participants. On hand, we found that users did not have enough discretion in selecting the partic of the community and buildings that they wished to comment on. In addition, ▮ constrained to the square shape of the grid: even if they only wanted to comme building in a corner of this large square, they had to select the entire square.

Menu of like and dislike

Input textbox

Submit

Clear

Zooming in and out

Back and forward

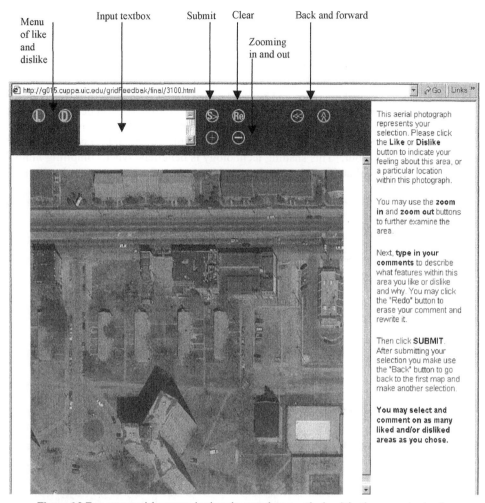

Figure 6.9 Two-way spatial communication: the nested map method and the three zooming levels.

6.5.2.2 Freehand

This project introduces an entirely new technology for enabling user feedback using Web-based maps. In this new prototype (http://g015.cuppa.uic.edu/zhaoxia/ blankcanvas/sketch.html) (Figure 6.10), participants can go to the online survey Web site and use a drawing tool to select the areas of the community that they wish to comment upon; their locational choices are not limited by the pre-defined geographic areas of the square grid. On the initial screen, the participants view a structure base map of the community along with two buttons labelled 'Click to select area with drawing tool' and 'Click to type in comments'. When a participant clicks on the drawing button, the cursor turns into an arrow and upon pressing and holding down the left button of the mouse, it starts to draw. The user may draw any shape on the map and when the mouse is released, the lines close on themselves to form a polygon. If the user does not draw an enclosed shape, the program approximates the line that closes the shape into a polygon.

Once the participant draws a shape and releases the mouse button, the immediately filled with a light colour. As the participant continues to select areas with the drawing tool, other shapes are filled in as well. When areas ov program indicates this by increasing the density of colour in these areas. A participants delineate most frequently will be the darkest, while the areas chosen would be lightest. This works through the placement of a very fine grid at the b map. When a participant delineates an area, cells underneath will be activated create one tone. As participants highlight additional areas, the overlapping are be activated twice, generating 'double' tones and hence increasing the dens colour. This shading technology is beneficial when compiling multiple users' r so that the darker shades would represent the most frequently chosen locations.

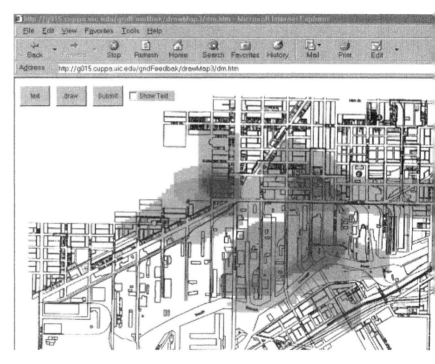

Figure 6.10 Two-way spatial communication: freehand sketch on the Web.
Colour gradation indicates intensity of concern.

Once the user has drawn and selected an area with the drawing tool, the p can then click on the text button to type in comments about the delineated area to avoid clutter on the map generated by multiple annotations when a participa several areas, we added a check box that would allow a participant to turn the t on or off. The participant may add a text box for each section of the commun selected. It is possible to make two separate interfaces: one for drawing an about liked locations and one for disliked locations. This may assist in sorting and negative opinions in the database. Another possibility would be to simply interface and allow participants to record all opinions, positive or negative.

The most important feature of this project is that we replace the Java applet freehand sketching capabilities using Java to create the Web interfaces so that t not restricted to pre-determined geographic areas within a grid. This new capability of doing freehand sketching on the Web may have numerous applications. People can make their own maps and then share these maps with other participants. Another possibility is that the map could be a photographic image and we could ask participants to comment on the scene by drawing directly on the picture, indicating elements that they like or dislike. Also, we hypothesise that community residents will appreciate this feature since it enhances the ability of the computer to mimic traditional public participation tools that people generally enjoy, such as drawing and writing comments on a paper map.

6.5.2.3 Toward a Complete Interactive Web Map Design

The prototypes discussed above each represent a particular feature; some are designed to sketch, others to type in comments. Some are grid based and others are based on freeform drawing. Some have zooming others do not. We are in the process of developing a Web site (http://g015.cuppa.uic.edu/mapTool3/mt.html) that will combine the best design features of the interfaces described above. Users will have options to work with or without the grid, to zoom in and out, to add layers, to type in comments, to sketch, and so on. This has apparent advantages (more choices) and disadvantages (confusion, technical difficulties to create the interface and to work out the database). This site is currently undergoing further development.

6.5.2.4 Compositional Methods

The final challenge we attempt to address using Web maps was how to allow participants to essentially create their own maps. This prototype (http://g015.cuppa.uic.edu/zx/sketch2/project3.htm) (Figure 6.11) was developed for a slightly different use than the ones described above. While the others use maps of the community to help residents share their opinion and knowledge about the community by selecting locations and typing comments, this prototype aims at allowing community participants to draw alternate site plans for a specific location, such as to show a preferred arrangement of houses in a new subdivision. Called the Collaborative Decision Support System, it utilises the freehand drawing technology described above. Users actually draw their own boxes (to represent houses, other buildings, or land use zones) and then move them around into a desired configuration. The purpose of this project is to allow users to create and compose maps online.

In the current prototype, the boxes are drawn on a structure base map of Pilsen. The primary buttons available are 'draw', 'move', and 'copy'. Users may correct their work with 'delete' and 'clear' buttons. When the drawing is complete, the participant clicks 'submit' to send the drawing to the server database. The user is then able to view all the drawings previously submitted by other users. While at present the prototype simply demonstrates the capability of creating and moving the boxes, there are many possible applications for the tool. It could be used for planning work on a small scale; for instance, the survey site could open with a site map and a description of the particular site along with a set of development plans. Users would be given a set of instructions, such as asking them how they would prefer to site the structures in a development that calls for 8 single family homes, 16 town homes, a community centre, and a playground. Users would then create the boxes and move them around on the screen until they developed a

satisfactory plan. Some of the fixed details of the site could already be indica
site map, such as the direction of traffic, existing driveways, etc.

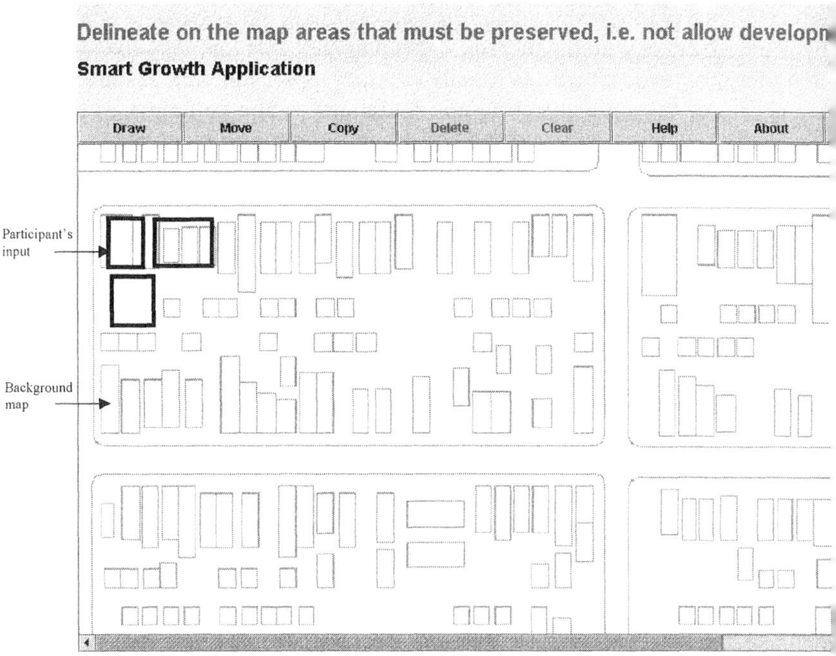

Figure 6.11 Two-way spatial communication: compositional method.

Alternatively, the proposed issue could be planning work at a larger scale,
land use and zoning plan. In this case, participants could help deal with the
urban sprawl by drawing shapes of different colours to indicate their prefer
placement of various types of land use zones. As the Web site is further develop
need to incorporate a legend to explain what each drawn shape is representin
various users could interpret one another's sketch plans.

6.5.3 Three-way Spatial Communication

Three-way communication enables users not only to view and input data but als
input data of all participants. The user is able to tap into the database and view a
The user may receive a return map in real-time showing the accumulative respo
participants.

In the following example, (http://g015.cuppa.uic.edu/gridFeedbak
18St/grid_18St.html) we created a 'real-time feedback' interactive map. Pa
could view a structure base map of Pilsen and choose areas of 'like' and 'disli
could also type in comments about the selected areas. Once their selecti
submitted, they could immediately receive a return map showing the acc
evaluative image of the examined area. This worked in the following way: t
servlet on the database is contacted and the data sent to the servlet translates
information into numeric, string data that is stored into the database. Once the d

updated the data servlet is called which queries the database for all the relevant information and then translates the data into visual information and sends it back to the user. The visual information includes the cumulative 'like' and 'dislike' opinions of the area, represented by increasing intensity of colour along with the user comments for each block in the grid. The data obtained from different surveys can be co-related as it is stored in the database, sorted by latitudes and longitude. The grid was a good structure that was easy to code and to be read by the Oracle database that created the return map.

Several of the other Web design interfaces can easily be upgraded to function as three-way communication mediums. For instance, in the Collaborative Decision Support System described above, we have added a feature whereby when users submit their design, they are then able to view all the drawings previously submitted by other users.

6.6   IMPLICATIONS FOR PARTICIPATORY PLANNING

Public participation in a community planning process is important in democratic society; however it is a complex undertaking. Our research explores state-of-the-art information technology (IT) to facilitate the process. Web-based mapping opens up the potential for involving a wider range of people by bridging time and space. At the same, IT has the potential to automatically collate participants' responses and ideas in a cohesive manner as described in the previous examples. Traditional public participation in planning usually relies on same-place and same-time meetings, which restricts involvement. Web-based mapping can be utilized for widening and diversifying channels of communication among the public, planners, and politicians.

Our research has asked the question, how can Web-based GIS aid participatory planning? It would be useful to place this question in the framework provided by Weidemann and Femers (1993). They presented a public participation ladder model that arranged the tasks of public participation in a vertical dimension. The order, from top to bottom, is as follows:

1.   public participation in final decision;
2.   public participation in assessing risks and recommending solutions;
3.   public participation in defining interest, actors and determining agenda;
4.   public right to object;
5.   informing the public; and
6.   public right to know.

Kingston (1998) argues that most Web-based GIS models are confined within the bottom two rungs of the public participation ladder; the 'public right to know' and 'informing the public'. Our research aims at expanding the role of Web-based GIS from being limited to the last two tasks to including as many as all six tasks. We envision that a loop of communication facilitated by two-way spatial communication may enable the public to 'object', to 'define interests and agenda', and to 'assess risks and recommend solutions'. This could be possible by further developing Web-based mapping that employs a wider range of increasingly powerful Web technology.

Present commercial Web-GIS software mainly enables one-way communication. The method is appropriate to allow the public to access spatial information from remote places such as home or office. Two-way communication on Web-based maps is intended to not only provide spatial information about a particular planning problem, but also to provide a forum for the public and planners to express their perspectives and concerns in

a spatial format. Planners as well as government officials can learn ab
knowledge, which is necessary for sound planning. The three-way comm
concept further supports democratic decision-making by allowing the public to
opinions of all participants.

The tools we have developed in these prototypical interfaces have the po
be used by a variety of agencies for multiple purposes. For example, it has been
that transportation planning and local 'comprehensive' planning tend to occur s
resulting in some cases in policies that work at cross purposes. A partial solut
be that comprehensive planning and transportation planning leaders could in
these kinds of Web-based tools to both inform the public and also learn abc
concerns and views. By sharing future planning ideas and learning abo
concerns, costly mistakes could be avoided. This kind of communication could t
address other issues such as urban sprawl, creating subdivisions, transportation
landscaping, and identifying advantageous options for development and envir
protection.

6.7 CONCLUSION

It is increasingly important to direct research to discovering the most effective
and tools for interacting with the public using maps. As there has been a
increase in the number of people exposed to and using screen-based g
information products for general use, two-way communication of spatial inf
must become more efficient and more easily comprehended. Researchers in
have estimated that 'up to 90% of all business data has a geographic compone
News, 1997), while an estimated 80 per cent of all government information is
referenced (Huxhold, 1991). Governments themselves publish geographic inf
on-line. For example, the San Diego, California Police Department publishes
crimes on a public Web site just 24 hours after the incidents occur. Local
agencies use the Web as an adjunct to the traditional public meeting format. A n
U.S. government agencies, including the U.S. Census Bureau (http://www.cens
the Geological Survey (http://www.usgs.gov), and the National Cancer
(http://web.nci. nih.gov/atlas), provide maps via the Web. Without direct
into which designs, tools and methods of presenting and receiving spatial inform
most easily used and comprehended by the public, we may be misinformed in e
the level of communication that is actually occurring.

In this chapter we have described a variety of digital map designs and tool
for a dual purpose: to communicate spatial information to average map user
allow those users to navigate and make selections on the maps in order to give
into a community planning process. We have provided examples of solutions to
problems of creating interactive screen-based maps, which include navigati
geographic area maps and making selections on maps. New Web technology ha
possible to create map-based surveys to receive feedback from the general public
not yet clear what kind of graphic design alternatives and digital map designs
are the most useful for novice map-readers. This research attempts to explore
interface designs with the purpose of finding which combinations of online t
maps are most productive in soliciting feedback in a community planning proce
future, we plan to empirically compare these interfaces to learn more about ho
comprehend screen-based maps vs. paper-based maps. More research is needed

to understand the unique challenges and important advantages of Web-based maps in general, and the usefulness of each graphic design and tool in particular.

Our project has explored how Web-based maps can be advanced beyond mere information provision to actual two-way interaction with the public. Tools such as freehand drawing may provide new avenues for people to take a greater role in public decision making. By beginning to examine the graphic design alternatives possible on the Web, we are taking one step further toward understanding how people comprehend and utilize screen-based maps.

6.8 REFERENCES

Aitken, S. and Michel, S., 1995, Who contrives the "Real" in GIS? Geographic information, planning and critical theory. *Cartography and Geographic Information Systems*, **22** (1).

Al-Kodmany, K., 2000, Extending geographic information systems to meet neighbourhood planning needs: The case of three Chicago communities. *The Journal of Urban and Regional Information Systems Association,* **12** (3).

ARCNews, 1997, Visio Corporation adopts MapObjects. *ARCNews*, Environmental Systems Research Institute, Inc. **19**(2), p. 1.

Chicago Fact Book Consortium (CFBC), 1990, *Local Community Fact Book Chicago Metropolitan Area*. (Chicago: Academy Chicago Publishers).

Craig, W.J., 1998, The Internet aids community participation in the planning process. *Computers, Environment and Urban Systems*, **22**(4), pp. 393-404.

Curry, M., 1995, Rethinking rights and responsibilities in geographic information systems: Beyond the power of the image. *Cartography and Geographic Information Systems*, **22** (1).

Edwards, C., 1984, *Overlay Drafting Systems*. (New York: McGraw Hill).

Hopkins, L., 1975, *Optimum-Seeking Models for Design of Suburban Landuse Plans*. Unpublished Dissertation.

Huxhold, W.E., 1991, *An Introduction to Urban Geographic Information Systems*. (New York: Oxford University Press).

Kim, D., 2001, *Design and Assessment of Web-based Geographic Information Systems*. Masters project, University of Illinois at Chicago, College of Urban Planning and Public Affairs.

Kingston, R., 1998, Web based GIS for public participation decision making in the UK. A paper presented at the *Empowerment, Marginalisation, and Public Participation GIS* Meeting in Santa Barbara, California October 14-17th, 1998.

Krygier, J., 1999, *Public Participation Visualization: Conceptual and Applied Research Issues,* http://www.geog.buffalo.edu/~jkrygier/krygier_html/

Lynch, K., 1960, *The Image of the City*. (Cambridge: MIT Press).

McHarg, I., 1969, *Design with Nature*. (New York: Natural History Press).

Nasar, J., 1998, *The Evaluative Image of the City*. (London: SAGE Publications).

Obermeyer, N., 1995, The hidden GIS technocracy. *Cartography and Geographic Information Systems*, **22**(1).

Obermeyer, N. and Pinto, J., 1994, *Managing Geographic Information Systems*. (New York: Guildford Press).

Peng, Z-R., 1999, An assessment framework for the development of Internet GIS. *Environment and Planning B: Planning and Design* **22**, pp. 117-132.

Rundstrom, R., 1995, GIS, indigenous peoples and epistemological *Cartography and Geographic Information Systems*, **22**(1).

Sanoff, H. 1991, *Visual Research Methods in Design.* (New York: Van Reinhold).

Shiffer, M., 1995, Interactive multimedia planning support: Moving from si systems to the world wide web, *Environment and Planning B: Planning a* **22**.

Shiffer, M., 1998, Multimedia GIS for planning support and public *Cartography and Geographic Information Systems*, **25** pp. 89-94.

Weidemann, P. and Susanne, F., 1993, Public participation in waste ma decision making: Analysis and management of conflicts. *Journal of F Materials*, **33**.

6.8.1 URLs: Examples and References

`http://www.mluri.sari.ac.uk:80/~mi550/landscape/`, Macaulay Land Use Institute (Web survey using images).

`http://www.ccg.leeds.ac.uk/vdmisp/`, Kingston's Web-based decisic systems in Britain

`http://www.uic.edu/cuppa/gci/about/index.htm`, University of Illinois at ((UIC) *Great Cities Institute*

`http://www.uic.edu/cuppa/udv/`, UIC's *Great Cities Urban Data Vis program* (GCUDV)

`http://www.census.gov`, The U.S. Census Bureau

`http://www.usgs.gov`, The Geological Survey and

`http://www.nci.nih.gov/atlas`, The National Cancer Institute

6.8.2 URLs: Project Interfaces

`http://www.uic.edu/~kheir/layer/p1`, *Map Overlay.*

`http://e036.cuppa.uic.edu/arcims/north-lawndale`, *ArcIMS for North Law*

`http://e036.cuppa.uic.edu/arcviewims/north-lawndale/index.html`, *Arc) for North Lawndale.*

`http://g015.cuppa.uic.edu/gridFeedbak/final/index.html`, *Nested Maps.*

`http://g015.cuppa.uic.edu/feng/project.html`, *Thumbnails.*

`http://g015.cuppa.uic.edu/gridfeedbak/xarial18St/arial_18St.html`, *G Map-Survey Interface.*

`http://g015.cuppa.uic.edu/gridFeedbak/drawMap3/dm.htm`, *Real-Time Interactive Map-Survey.*

`http://g015.cuppa.uic.edu/zhaoxia/blankcanvas/sketch.html`, *Freehand and Selection Tools.*

`http://g015.cuppa.uic.edu/mapTool3/mt.html`, *Integrated Interface with (Features.*

`http://g015.cuppa.uic.edu/zx/sketch2/project3.htm`, *Collaborative Support System* (composing site plans online).

7

A collaborative three dimensional GIS for London: Phase 1 Woodberry Down

Andrew Hudson-Smith and Steve Evans

7.1 INTRODUCTION

In this chapter we outline the various issues involved in developing a three dimensional geographic information system (3dGIS) model for a large urban area, such as London. There are now many techniques which can be used to construct two-dimensional (2d) and three-dimensional (3d) models of cities and their characteristics, which are useful for various types of visualisation. Until quite recently the main approach was based on rendering wire frame models of buildings based on Computer-Aided Architectural Design (CAAD/CAD). However, with the widespread development of Geographic Information Systems (GIS), there has been a move from the 2d map to the 3d block model which is based on extruding building plots, street line data, and basic topography. At the same time there has been a move towards the distribution of information relating to urban form and its representation via the Internet. Research at CASA into Internet based urban planning techniques has resulted in techniques whereby the common CAD based models of urban form can be adapted into a networked collaborative design and information system. A system which can be queried, explored and indeed inhabited through standard desktop machines linked to the Internet via standard dial up connections is currently being piloted.

This chapter addresses two issues, which involve our proposal to construct a 3dGIS for London, firstly the construction of an Internet distributed three-dimensional model of London and secondly its integration with web enabled GIS. It explores the current 'state of the art' in 3d GIS and compares these with other photogrammetric techniques which produce 3d models. Through the use of multi-way interaction, it demonstrates how photorealistic models of the built environment can be distributed and queried online through networked GIS. The paper also demonstrates how such models can be populated, creating distributed virtual environments underpinned by a GIS. In short it details methods to communicate, distribute and query built form for the purpose of *Online Planning*. We will elaborate our proposed paper in three sections: spatial data technologies, model construction and dynamic 3dGIS for Online Planning.

7.2   SPATIAL DATA TECHNOLOGIES

For a model to be useful it must tie in with a range of consistent spatial data. W
distinction between *content* and *geometric* data. Content data is now widely av
different levels from the Census, the Office of National Statistics, the Greate
Authority, the Valuation Office, and a variety of property companies an
concerns. There are four sources which are being built in as layers and which
query and display:

- Public domain data ranging from population to employment, deprivation i
  and related socio-economic attributes;

- Detailed data sets that have been compiled for one-off projects such as i
  Cross Development etc;

- Unconventional content such as visual data, advertising data etc, which is
  and can be added in stages;

- Private data sets which we are soliciting as part of the wider support for tl
  will be sought on a case-by-case basis.

Geometric data for the basic visual content is a different matter. There are
of data providers and there is the possibility that we might produce our own data
effectiveness of these options is crucial to a successful model. However, this sp
is of little use if it is only available to those with the technical expertise or the
software. Access to the GIS must be easy and widespread. Even though the
underlying GIS have existed for over 30 years, it is only in the last 10 y
technological developments in computing have enabled planners to exploit the
(Coppock and Rhind, 1991). Consequently virtually all of the GIS systems imp
by the planning community have been since 1990 (Masser and Craglia, 199(
more recently GIS, like many related technologies, has become 'web-enabled'
the Internet does not change the fundamental nature of GIS (Harder, 1998), it (
the technology to an infinitely wider audience.

GIS is no longer only reserved for the technical experts or those with the
software. Through the use of the Internet, home users are increasingly exploiti
plan a journey or locate the nearest cash-point. In most cases people are explo
software without even realising they are using GIS, yet it provides a perfect int
to the technology which can be exploited by the planning community. As the u
technology becomes more widespread and more accessible to home users, it is i
that its role in planning will increase. If this spatial data and associated informati
taken one stage further and visualised in a 3d environment by home users, the
able to offer a system that may be even more appealing to architects and urban
alike.

7.3   CONSTRUCTING THE MODEL: PHOTOREALISM AND DISTRIB

The aim is to provide a detailed geometric mapping and three-dimensional visu
rendered to a sufficiently high level of accuracy to provide users with a definite
'virtual' London is 'real' London. This sense of location and place needs to be
while bearing in mind Brutzman's (1997) six components of Internet distribut
dimensional graphics. They are as follows: connectivity, content, interaction, ec
applications and personal impacts. The combination of these six components is
our understanding into how networked three-dimensional graphics, and the

virtual environments, are produced by the modeller, distributed over the Internet and browsed by the end user. Figure 7.1 illustrates the interlinking features of Brutzmans' six components with an additional seventh, that of file size. Each of these features and components are examined in turn in relation to creating and distributing a three-dimensional model of London.

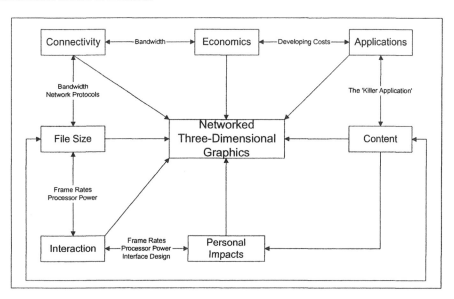

Figure 7.1 Interpretation of Brutzmans' six components of graphics internetworking with the addition of file size.

The first, and perhaps most important component is connectivity. Connectivity refers to the capacity, bandwidth, protocols and multicasting capabilities on the networking infrastructure (Rhyne, 2000). Although networking is considered to be 'different' than computer graphics, network considerations are integral to the production and distribution of large-scale interactive three-dimensional graphics. Graphics and networks are two interlocking halves of a greater whole: distributed virtual environments (Brutzman, 1997). Connectivity, in terms of available bandwidth, is a decreasing problem. At the start of research, in 1997, the average home users connection speed was 28.8bps with many users connecting via slower 14.4bps analogue modems. At present average connection speeds for home users are 56Kps (Graphics, Visualisation and Usability, 1998). The increase in available bandwidth has allowed an increase in file size for the virtual environments produced during the research into Online Planning. Table 1, lists theoretical download times, based on a perfect connection, of a typical 1.5mb file for a virtual environment. Download time is listed in relation to connection speeds and thus increasing bandwidth availability.

Table 7.1 Download times according to connection, adapted from Telegraphy, 2000.

33.6Kbps	56 Kbps	128KBps ISDN	784 Kbps xDSL	5Mpbs Cable
7.5 minutes	4 minutes	1.5 minutes	16 seconds	2.5 seconds

Download times are representative of distribution mediums using a cli model, whereby the entire file is required to be downloaded before tl environment can be rendered. Alternatives to the client/server model are discu There is a direct relationship between connectivity and file size in the client/serv This relationship represents a bottleneck in the ability to distribute three-di models online.

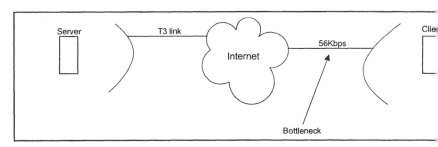

Figure 7.2 Bottlenecks in access speed, adapted from Halabi, 1997.

As Figure 7.2 illustrates, even if a model is hosted via a T3 link tl connection the client will obtain is 56Kbps using a 56Kbps standard moder three-dimensional environments therefore need to be tailored to the average c rate, currently 56Kbps. As already stated, bandwidth is a diminishing problem, c increasing as cable modem and Digital Subscriber Line (DSL) technologies are i to enter the home. However, at the current time the successful distribution environments is inexorably tied to the ability to distribute within available b. thus file size is all-important.

Linking connectivity and file size is interaction. Interaction involves a presence and the ability to both access and modify content; it also defines th location and place in a three-dimensional scene. The level of interaction is a k to consider in the modelling of the built environment. Despite the ability of r three-dimensional graphics to create imaginative and complex environments th often limited by low levels of interactivity, either due to system limitations or design.

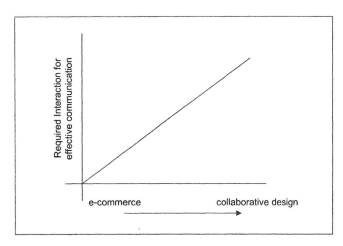

Figure 7.3 Level of required interaction with application.

The level of interaction is tied in with the content and application (Figure 7.3). E-commerce requires a lower level of interaction than, for instance, collaborative design. In e-commerce applications the user normally only requires the ability to view the product from a range of angles. Collaborative design requires a more sophisticated level of interaction with the ability to walk and flythrough the environment in addition to the ability to move and add and subtract objects. E-commerce is a recurring theme in the development of three-dimensional graphics as it is the driving force behind the development of both web based 3d software and the form of distribution. E-Commerce is therefore a limiting factor on the development of collaborative design.

Personal impacts are linked with both the level of interaction and the content of the environment. It can be seen as the sense of wonder or frustration when interacting and navigating a virtual environment. Interlinking these components is interface design. How the user navigates and explores the environment is crucial to the level of wonder or frustration experienced.

Content can be seen as any information, dataset or stream that is transported via a networking protocol, in our case models of the built environment. Content is an emotive issue in the modelling and distribution of three-dimensional models as despite all the hype, there is yet to be a 'killer application' for online three-dimensional graphics. Currently three main content driven applications can be seen, these are multi-user gaming, shared virtual environments and e-commerce. These 3 applications are shaping the delivery of virtual environments.

Economics is all-important, if three-dimensional graphics are to become commonplace on the Internet, entry requirements need to be at a consumer cost base. This has to include the cost of the whole package, from cost of connection and hardware, to the cost of viewing and development software. Economics is especially important in terms of use of three-dimensional graphics by local authorities.

Three-dimensional communication of the built environment, in order to communicate a sense of location and place online, is dependent of Brutzmans' six factors and a seventh, file size. For effective communication utilising networked three-dimensional graphics a rethink of traditional modelling and distribution techniques is required.

The modelling of urban scenes as a portrayal of the existing environment is crucial to view any proposed developments in context and portray a sense of location and place to the use. It is the modelling of the existing environment that poses the most difficulty, as Kjems (1999) states. It is much more difficult to build a 3d model of an existing environment than a new development. With this in mind methods are examined based on four levels of detail and abstraction, namely panoramic visualisation, prismatic primitive, prismatic with roof detail and full architectural. Firstly we examine panoramic visualisation.

Panoramic visualisation is not three dimensional, in that it consists of a series of photographs or computer rendered views stitched together to create a seamless image. Rigg (1998) defines a panorama as an unusually wide picture that shows at least as much width-ways as the eye is capable of seeing, if not a greater left-to-right view than we can see (*i.e.* it shows behind you as well as in front). Figure 7.4 illustrates a sample panoramic image of Canary Wharf Square in London Docklands.

Although panoramic images are two dimensional, as they are constructed from a series of photographs, the effect is considerable realism (Cohen, 2000). Panoramic images are not a new concept brought about by the rise of the digital age, indeed they have been made since the 1840's with the introduction of the first dedicated panoramic cameras. However, it was not until 1994 and the introduction of QuickTime Virtual Reality (QTVR) for the Apple Macintosh that panoramic production, from the stitching of

a number of photographs, became available on consumer computers for the
QTVR works by taking a sequence of overlapping images and automatically ali
blending them together to create a seamless panorama. The resulting pic
photorealistic capture of a scene taken over the time required to capture th
typically between 30 seconds and two minutes. Panoramas are viewed online v
plugin or Java applet. The viewer renders a section of scene allowing the user t
zoom the panorama using a combination of the mouse and keyboard. Ea
panorama is defined as a node, and hotlinking between a series of panoramas
multi nodal tour.

Figure 7.4 Panoramic image of Canary Wharf Square, London Docklands.

In terms of Brutzmans' components of networked three-dimensional
panoramas score highly on use of available bandwidth and file size. To view an
a panoramic image all that is required is the plugin or Java applet and the ir
Based on a medium level of compression, image file size for a typical panorami
150k, or 200k including the Java viewing applet. Capture techniques are both
low cost with stitching software typically available for under £100. Panoramic i
thus well suited for the communication of existing locations via the Internet,
photorealistic representations with low files size. Interaction is however limited,
pan and zoom or link to HTML documents but as the image is two-dimensional
level of interaction is possible. The use of panoramas becomes more problema
developments are to be visualised. This involves integrating a three-dimensio
with an existing panorama or creating a rendered photomontage, essentially au
reality. Augmented reality has been used for the purpose of Online Planni
example of the Bridge Lane Planning Enquiry, more details can be found
http://www.casa.ucl.ac.uk/olp/.

For the production of full three-dimensional models of the exist
environment there are three critical factors - building footprints, roof morph
height data. It is the combination of these factors that allows the creation o
models. Building footprints are widely available in the UK, most commonly ir
of Landline data from the Ordnance Survey. The data is however both pro
expensive and detailed for online usage.

Figure 7.5 illustrates building outlines derived from Landline data for a section of Central London. At a cost landline data can be obtained, however the main problem is the acquisition of suitable height data and roof morphology.

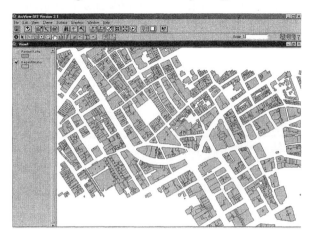

Figure 7.5 Landline derived building outlines in ArcView.

Average height data can be purchased 'off the shelf' from mapping companies such as Cities Revealed. This data provides the average height according to building footprints, problems have however been experienced with this data. Average height is used, calculated over enclosed building footprints. This is not detailed enough to produce a convincing three-dimensional model. In addition the fact that it is based on extruded footprints results in a high level of detail in terms of final polygon count and thus in relation to Brutzman (1997) makes the data unsuitable for web distribution. Comprehensive data can also be obtained from range imaging methods, the most widely used being Light Detection and Ranging (LIDAR). LIDAR provides a high-resolution three-dimensional surface, which can be imported into a GIS and draped with an aerial photograph (Figure 7.6). LIDAR is at the high end of the data range scale and as such is not suitable for the production of models aimed at online distribution, although combined with building footprints average height can be extracted from the LIDAR dataset.

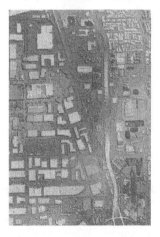

Figure 7.6 LIDAR-based City Models (http://www.globalgeodata.com).

Figure 7.7 illustrates a section of Central London extruded from building
up to an average height derived from LIDAR data. The resulting model is a
representation of Central London. The model is both crude and unwieldy ir
required processing power and file size. Manageability of the model can be imp
considerable generalisation of the base data is required. Generalisation of ba
problematic.

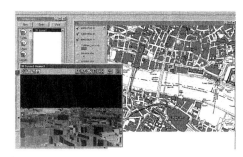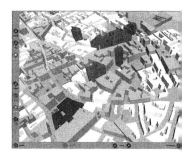

Figure 7.7 Extrusion by average height derived from LIDAR in ArcView.

Prismatic models lack any significant architectural detail and thus do n
any compelling sense of the nature of the environment (Batty and Smith *et c*
Roof morphology can be added either within a GIS or via a standard modellin;
such as Microstation or 3D Studio. There has been considerable research e
recent years into developing the capabilities of GIS to handle three-di
information of the built environment (Faust 1995). This has often been achieve
the linkage of CAD technologies to a GIS database (Ligget, Friedman and Jeps
Linking GIS to CAD is however a tentative process. Figure 7.8 illustrates the
PAVAN a modelling package for the MapInfo GIS package.

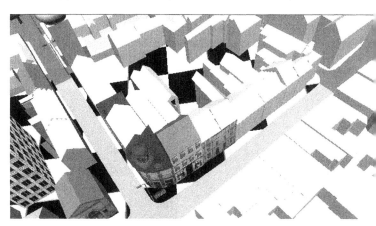

Figure 7.8 PAVAN output from MapInfo, illustrating Roof Morphology.

PAVAN enables roof morphology and texture maps to be added to height
While this is adequate for basic modelling, the level of realism is low and it
knowledge of the modelled areas roof structure, data which is not commonly
without a comprehensive area survey. Where roofing morphology is not kno▾
method for modelling is required.

Methods to rapidly extract and texture map both building outlines and roof morphology have become readily available in the last 18 months. A result of the increase in personal computing power and the demand for realism, predominantly in gaming environments, has given rise to packages such as Canoma from Metacreations, GeoMetra from AEA Technologies and Image Modeller from RealViz. These packages, aimed at creating models optimised, in terms of file size, while retaining a high degree of realism, are directly suited to the production of models aimed for distribution on-line and fall within the requirements of Brutzman (1997).

The following example provides an illustrative walk through of creating a texture mapped three-dimensional model, of a typical new build development in the UK, using Canoma from Metacreations. Canoma is typical of the new range of low cost photogrammetric modelling packages.

Figure 7.9 Canoma modelling stage 1.

The model was constructed from two photographs, taken with a Nikon CoolPix 850 digital camera, Figure 7.9. The photographs were framed to ensure that all four corners, and any shared geometric features of the building, were in view. The first stage to constructing the model is mentally dividing up the building into a series of primitives, these primitives will be aligned, joined and stacked to create a wireframe version of the building. Once the building has been divided into basic shapes the first primitive can be placed, in this case a box which constitutes the main area of the building. The correct placement of the first primitive is all-important. From the first primitive Canoma calculates the position of the ground plane and the camera position. Pinning the corners to the corresponding points in each photograph places the box, the pins are represented as triangles, Figure 7.9. Where a corner is not visible, as is the case in the bottom right photograph, a bead, a round dot, is placed to guide the primitive to roughly the correct position.

Stage 2 creates the central roof structure, this is created using a 'stack' placing the selected roof shape primitive directly on top of the first box. A comb pins and beads are used to align the primitive with the actual photographs (Figur

Figure 7.10 Canoma modelling stage 2.

The third stage repeats the procedure of creating the first box primitive and to create the front section of the house. To ensure the new section of the correctly aligned it is 'glued' to the first box primitive (Figure 7.11).

Figure 7.11 Canoma modelling stage 3.

Figure 7.12 Canoma modelling stage 4.

The wireframe is now taking shape and matching the two photographs, the front porch section and chimney are added in the same manner as stages 2 and 3, using a combination of pins, beads and glue, Figure 7.12. The model can now be automatically texture mapped and exported in the desired distribution file format. The example provided is for a single building, whereby a couple of images are sufficient to create the three-dimensional model. Two images are sufficient for the wire frame model as there a number of linked geometric reference points in each image, the model is also made up of basic standard primitives. For more complicated large-scale urban areas the addition of oblique aerial photography is required to provide an overview of the entire scene. Combined with street level views urban scenes can be constructed, Plate 5 illustrates a model of Canary Wharf, London, modelled in Canoma. The model was produced using a combination of aerial photography and streetlevel photographs taken from the Canary Wharf Square panoramic example (Figure 7.4).

Once a scene is constructed, all-important to its successful placement online is the file format it is saved in and the resulting format used for distribution. The format chosen is a critical factor in the balance of Brutzman's components for networked 3-dimensional graphics. Our preference is for a combination of Virtual Reality Modelling Language 2.0 (VRML), an open source language and Viewpoint from Metacreations. It is not intended to go into detail on distribution formats here, a full comparison of web3d file formats can be found online at http://3dgraphics.about.com/compute/3dgraphics/.

7.4   DYNAMIC 3DGIS FOR ONLINE PLANNING: WOODBERRY DOWN

We have established that to place three-dimensional models online the traditional '3D GIS' route is not suitable, what is required is a low bandwidth compliant model produced from the emerging range of photogrammetric software. This therefore rules out the common route of developing the 3d model from within the GIS itself and integrating the three dimensional and spatial data in one system.

It is more feasible to link a web enabled GIS or its Internet Map Server with web enabled 3d modelling systems. This can be done by developing unique database key fields that can address match a location in the GIS with an associated building or location in the 3d model. If the user chooses to move within the 3d model then this movement is detected and the location in the map interface updated, and vice versa. In addition a wealth of other data can be stored 'beneath' the map interface. Users may only view a photorealistic 3d model of a building, but the map can deliver a wealth of information about the building or the surrounding area. This could include air pollution levels for the area, the name and address of the building, details of the history of the building or even a link out of the system to the website of the architect's practice who built it.

We have developed a 3dGIS in the London borough of Hackney for collaborative planning in the redevelopment of the Woodberry Down estate. The system is based on the ideal of opening up the redevelopment system to the local residents, via the internet, through an interface which gathers together communication and visualisation techniques aimed at facilitating local democracy in the planning process. Funded by the Woodberry Down Redevelopment team and supported by the Architecture Association and Hackney Building Exploratory, the project demonstrates how in a short period of time, 3 months, it is possible to implement a collaborative decision making system, running in people's own homes. One of the issues arising from previous work into this field, Kingston *et al.* (1999), is that people do not have access to the tools that facilitate the communication, in our case the Internet. This has been remedied by securing funding to 'wire' 30 resident representatives (one, from each block of the flats in the redevelopment area) with free

Internet access and a Pentium III 600Mhz personal computer. This has ensure
interface is tailored directly to its users, allowing us to take into account Brutzm
components with a set of known limitations. The interface breaks down to the
components:

7.4.1 Arc Internet Map Server

ArcIMS, ESRI's latest generation of Internet Map Server has been used to de
GIS interface to the Information System. On a basic level this needs to simply
navigation tool around the area and a data structure to ensure a tight coupling be
3d and 2d elements of the system. However, on a more detailed level this can k
a search engine (for street names and postcodes), an information retrieval sy
example for discovering air pollution levels in an area or information about a
building or block of flats) and finally as a way of allowing the user to interacti
location geo-referenced comments about the area.

In order to avoid plug-in download times, the system has been restric
HTML site with the standard ESRI Java applets to provide improved functior
example rubber band selection of objects). The interface has then been custo
adapting the JavaScript files provided by ESRI in order to adapt and extend so
functionality of the system.

7.4.2 3D Models

Models of the area have been created in Canoma and 3Dstudio Max (versio
ensure a photorealisitc representation of the area while ensuring compliance
restrictions of a low bandwidth connection. Plate 6 illustrates a model o:
Gardens, Woodberry Down. The model is developed to be interactive, allowing
scenarios to be visualised in real-time by the local residents. These scenarios
place in association with the architects and developers contracted to redevelop
The redevelopment of Woodberry Down is a long term project, to ensure that tl
remain compatible with emerging web based technologies they are stored in
Max, an industry standard modelling package. Each model is linked to the
allowing each section to be displayed according to any of the parameters contai
GIS.

7.4.3 Panoramic Visualisation

The redevelopment area has been captured in a series of interlinked (via the
panoramas. Panoramas are an integral part of the site visualisation as they ɪ
360x360 degree documentation of the area as is at the time of capture. As wiᵗ
models they can be used to visualise future development in local context by au
reality.

7.4.4 Bulletin Board

Central to the system is the community-based bulletin board. The board alle
residents to log in and express their views on each stage of the redevelopm
enables the redevelopment process to move away from the standard 'village hall
scenarios in which only a few residents are able to have their say. Using a

coupled with free Internet access, all the residents are able to participate and have a fair say in the proceedings. The bulletin board, is not however restricted to the issue of redevelopment. It also includes sections on local/community news, a trading sections for good and general discussion. A range of subjects underpins the project, it is not only about the development, it is about the community as a whole, facilitating not only democracy but also community ties and relations over the Internet.

All the separate elements are in place and it is intended that the work will go live by mid 2001, coinciding with the residents being 'wired'. This will complete the picture of an Online Planning system for the redevelopment of Woodberry Down.

This project is phase one of a larger scale project – a Virtual London. We have established that large-scale models can be produced utilising a level of realism to ensure that a strong sense of location and place can be achieved. These models can be linked to spatial data via an Internet Map Server system which in turn can be linked to a bulletin board to facilitate communication. Updates on the work may be found at `http://www.onlineplanning.org` and `http://www.casa.ucl.ac.uk`.

7.5 FUTURE DEVELOPMENTS

Dynamic two-way interaction between the 3d models and ArcIMS is possible through the use of XML. Both Viewpoint (the format of the models) and ArcIMS operate through XML, enabling common 'tags' to be produced and thus enabling closer integration. It is also feasible to run the Woodberry Down project on hand held devices, such as Palm Pilots, while maintaining the required level of interaction. GIS, in the form of ArcPad, can be linked with both panoramic images and 3d models optimised to run on either Windows CE or Palm operating systems, leading the way to portable on-site visualisation and communication.

7.6 REFERENCES

Batty, M., and Smith, A., 2001, Virtuality and cities: Definitions, geographies, designs. In *Virtual Reality in Geography*, edited by Fisher, P.F. and Unwin, D.B. (London: Taylor & Francis).

Brutzman D., 1997, Graphics internetworking: Bottlenecks and breakthroughs. Chapter 4, *Digital Illusions*, Clark Dodsworth (ed.) (Reading, Ma: Addison-Wesley), pp. 61-97.

Cohen G., 2000, *Communication and Design with the Internet* (New York: W.W. Norton & Company), pp. 128-129.

Coppock, J.T., Rhind, D.W., 1991, The History of GIS. In *Geographical Information Systems: Principles and Applications*, edited by Maguire, D.J., Goodchild, M.F, Rhind, D. W., (Harlow: Longmans), pp. 21-43.

Faust, N.L., 1995, The Virtual Reality of GIS, *Environment and Planning B: Planning and Design*, **22**, pp. 257-268.

Graphics, Visualisation and Usability Center, 1998, 10th WWW User Survey, Technology Demographics, `http://www.gvu.gatech.edu/user_surveys/survey-1998-10/graphs/graphs.html#technology`.

Harder, C., 1998, *Serving Maps on the Internet: Geographic Information on the World Wide Web*. ESRI Press. pp. 16-24.

Kingston, R., Carver, S., Evans, A., Turton, I., 1999, A GIS for the public: Enhancing participation in local decision making. *Proceedings GISRUK 1999*. `http://www.geog.leeds.ac.uk/papers/99-7/`.

Kjems, E., 1999, Creating 3D-models for the purpose of planning. *Computers Planning and Urban Management,* `http://www.iuav.unive.it/strater pdf/C6.pdf.`

Liggett, R., Friedman, S. and Jepson, W., 1995, Interactive design/decision m. virtual urban world: Visual simulation and GIS. *Proceedings of the 1995 I Conference,* pp. 308-335

Masser, I., Campbell, H. and Craglia, M., (editors), 1996, *GIS Diffusion: The and Use of GIS in Local Government in Europe.* (London: Taylor & Franci: 110.

Rigg, J., 1998, *What is a Panorama? PanoGuide,* `http://www.panogu reference/panorama.html.`

Rhyne, T.M., 2000, *Internetworked 3D Computer Graphics: Overcoming B and Supporting Collaboration II.* Overview of 3D Interactive Graphics Internet, SIGGRAPH 2000 Course #28 Notes, `http://www.siggr %7Erhyne/com97/netgraf3/overview-netgraf3.html.`

Telegeography, Inc., 2000, *Telegeography 2000 Global Telecommuncatio: Statistics and Commentary,* Edited by G.C. Staple, (Washington: Telegeogra

8

Historic time horizons in GIS: East of England historic landscape assessment

Lynn Dyson-Bruce

8.1 INTRODUCTION

Historic Landscape Assessment (HLA) is a new methodology being developed to assess the historic dimension of the landscape which will enable the 'time-depth' and complexity of the landscape to be assessed. The HLA process charts the surviving, visible historic components within our landscape, to enable a better understanding of the processes of landscape genesis.

As a GIS project, HLA pays particular attention to metadata which documents the sources and decisions made in interpreting each polygon. The methodology uses a succinct set of attributes, within a simple database, to create a single layer. Within this layer are recorded all components that share common and/or diverse histories to create a complete coverage of the project area.

The metadata focuses on specific historic time horizons in conjunction with landscape types. This may enable the development of spatio-temporal models of landscape change, which could inform the decision making process to enable appropriate management of what is increasingly becoming recognised as a finite resource. This paper reviews the techniques developed for a project spanning six counties in the east of England.

8.2 BACKGROUND

Historic Landscape Assessment or Characterisation (HLA/C), is a broad-brush approach, which has been developed to identify and map the historic 'time depth' of the landscape. This approach is currently being applied across England on a county-by-county basis as part of a national programme, by English Heritage (EH), the major funding body, in conjunction with regional authorities. Work is in various stages of progress across the country.

HLA has developed from the traditional, 'paper based' methodology of Landscape Character Assessment (LCA) (Countryside Commission 1993). LCA is a methodology

devised to assess the character of regions within the country, on regional and lo
and has been applied on a broad scale (usually 1:50,000), across the country on
by-county basis. LCA has traditionally been a paper-based approach, but is
increasingly GIS based. However for HLA, each county has devised
methodology, either paper-based and/or GIS-based, but without being couched
national GIS or HLA methodological framework. It is therefore felt that res
collated on a national scale could reflect not only historic landscape diversity
methodological differences.

This paper will discuss and illustrate various issues including the method
metadata, which has been applied within the project, and its importance within
a proposal for a national series of historic landscape character types to be e
based on GIS national standards, to enable consistent comparative analy
combination of GIS and historic analysis of the landscape may help to achie
understanding of our cultural landscapes and their dynamics, and may assist ir
appropriate management strategies.

The methodology within this regional project has been re-defined to
account the basic principles of GIS, and to consistently reflect landscape cc
across regions. The result will be an apolitical seamless map, which will enable
project has been completed, spatial and Digital Terrain Models (DTMs) to be
and generated, as required of historic landscape patterning regardless of these
boundaries. In consequence results should reflect historic landscape diversity r
methodology.

Traditionally archaeologists have used a 'site-based' approach
archaeological sites and features are viewed as discrete entities within the l
which has now been recognised as having severe limitations for broad landscape
These records may be held within regional, or county based Sites and M
Records (SMR) or the National Monuments Records (NMR) managed by
Commission on the Ancient and Historical Monuments of Scotland (RCA
Scotland, or English Heritage in England, or CADW in Wales. This data is
represented as point data, albeit increasingly within a GIS.

It is not only difficult but would be inappropriate to extrapolate this point
the surrounding landscapes for wider landscape interpretation or assessment, as
no longer be historically contemporary with its surrounding landscape. Du
traditional 'site specific' approach, the weighting is on the 'site' rather than
context the 'landscape', which until now has been difficult to assess. In add
implies importance to the site, so that by default the surrounding areas or lands
of no, or of limited historical significance, and are therefore less important. Th
recognition and subsequently appropriate protection, makes the landsca
vulnerable to future development. However it has become increasingly clea
landscape has an historic dimension, which needs to be adequately assessed
management purposes, especially in response to current issues regarding susta
The landscape is an historic record in itself, also a finite resource, which is bein
or lost in certain areas beyond all recognition due to modern pressures, *e.g.* dev
mineral extraction, farming.

Before we can devise appropriate management strategies, we must first
resource in an appropriate manner. HLA seeks to redress this imbalance by a
methodology for assessing the landscape in terms of its historical developme
which this spot-site data may then nest and be assessed as to its contemporaneity

8.3   EAST OF ENGLAND PROJECT

8.3.1 Project Area and Remit

The East of England project was established to assess the whole of a broader administrative region, comprising the six shire counties of Bedfordshire, Cambridgeshire, Essex, Hertfordshire, Norfolk and Suffolk. This was considered to be sufficiently large and diverse in terms of its geography and historical development for meaningful analysis. A key objective of the project was to aid in the conservation of the historic environment by assessing the historic development of the landscape and using this information to inform its most appropriate long-term management.

8.3.2 Landscape History

In assessing landscape, in this case 'landscape history', it is necessary to be aware of certain forces which determine how our landscape has been formed. These include natural and anthropogenic forces, which are interdependent upon each other. In addition technological events and processes, in conjunction with socio-political reforms and policies, may have far reaching effects upon the landscape, and may be historically specific and therefore recordable. Humans have manipulated the naturally occurring landscape, producing dramatic and long lasting impacts on the landscape through time. Therefore it must be recognised that the landscape is an historic record in itself, which requires to be appropriately understood.

We are aware of the variety of landscapes around us, but few realise the 'time depth' of these landscapes, whether they have changed, and if so, how, the speed of change, or what frequency and how they relate to each other and to other determinants. The landscape is dynamic being in a constant state of flux.

HLA charts the surviving, visible historic components within our landscape. Results so far indicate that the history of the landscape is dynamic, diverse, complex and unpredictable. So far it has been shown that there are areas of continuity, discontinuity, contiguity, rarity, and diversity. The only consistent result is this diversity and dynamism.

8.3.3 National, Regional, Local Landscape Type Series

It is proposed that by using a variety of scales of approach we may begin to identify National, Regional and Local landscape types and thereby recognise areas of diversity and rarity, at appropriate scales. Some events are strictly local and do not need to be preconceived in a national taxonomy, but will fit into a broader grouping at the regional and national levels.

Therefore it is proposed that there should be a 'national type series' of historic landscape types into which all areas of study could feed, at the appropriate scale (see Figure 8.1). This need not deflect from other approaches and developments, nor each area's individuality, they only require prior thought and a linking field. It is proposed that this series should act as an 'umbrella' of generic types into which each region's diversity and uniqueness may contribute and support. This would result in 'national', 'regional' and 'local' types, a classification reflecting at each appropriate scale landscape diversity and detail, in historic terms. This would be taking the traditional 'grandparent', 'parent', 'child' model.

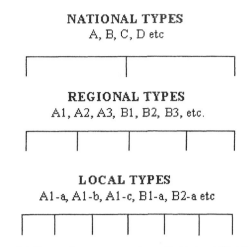

NATIONAL TYPES
A, B, C, D etc

REGIONAL TYPES
A1, A2, A3, B1, B2, B3, etc.

LOCAL TYPES
A1-a, A1-b, A1-c, B1-a, B2-a etc

Figure 8.1. Proposed concept as a schema for National HLA Types.

For example, at the national level one could have broad categories r‹ period, *e.g.* 19th century, 20th century, with generic landscape type, *e.g.* field woodlands, informal parklands. At the regional level these may be more subdivi‹ greater range of more specific landscape types, *e.g.* irregular fields systems fields systems, with more refined time horizons. At the local level one may ‹ specific landscape types reflecting local variations, with very specific historic *e.g.* Tithe, Enclosure or Estate maps cited. The complex palimpsest may only be the local/regional level. However, with a suite of attributes one could select required, and group appropriately for analysis at alternative scales, for combining specific historic time horizons into broader categories, or group‹ landscape types together – hence taking local variants and combining them int‹ regional groups.

8.4   DEVELOPMENT OF HLA

8.4.1 Background

The HLA methodology has developed from the seminal work carried out in (Herring, 1998), which mapped the landscapes according to 'Historic Character paper-based exercise, derived from the traditional LCA approach (C‹ Commission, 1993). It was the first attempt at assessing the landscape, rather t‹ data in historic terms.

Work in Scotland further developed the approach by mapping 'Historic using a Geographic Information System (GIS) (Dyson-Bruce, 1998: Dyson-Br‹ 1999), by Historic Scotland and RCAHMS, now centrally managed and a‹ RCAHMS. This ensures consistency of application and methodology across Wales has defined a 'Register of Landscapes' of specific or outstanding interest 1998) which ranks and values specific selected areas, but not the whole landscap‹

EH has used a wide variety of paper or increasingly GIS based methodo‹ determine 'Historic Landscape Character' (HLC) in different counties (Fairclou‹

across England. EH seeks to assess the entire landscape as an historical resource, rather than place import or rank specific areas. This is a new, dynamic approach still under development, especially as a GIS application.

8.4.2 Time Horizons within GIS

However it is the advent of GIS that has facilitated this approach, as the information may be represented in four dimensions:

- 2nd Dimension   horizontal space, *i.e.* N/S – E/W
- 3rd Dimension   vertical depth, *i.e.* topography
- 4th Dimension   time, *i.e.* time-depth or palimpsests

The horizontal, 2nd dimension definition of landscape components forms the backdrop of the mapping project. The topographic or elevation aspect, is secondary and may be derived later, using for example Ordnance Survey (OS) contour data.

The key element of the methodology is the treatment of historic events within the landscape as a series of time-horizons to enable spatial-temporal modelling, the 4th dimension. This together with the large quantity of complex spatial information is ideally suited to the GIS platform, which has enabled HLA to become a powerful tool. As a consequence the original paper-based methodology had to be adapted in response to the demands of various issues including GIS, technology, landscape complexity and archaeology. Therefore the GIS provides support to the HLA by including self-documentation regarding derivation of the data sources within the metadata, thus providing a clear audit trail of the process of assessment and analysis.

Earlier approaches were based on established Landscape Character Assessment (LCA) techniques and methodology (Fairclough, 1999). These were paper-based, which led, by their inherent limitations, to the creation of a single map or series of simplistic maps, achieved by the aggregation of data to create thematic historic landscape types. Paper-based HLA is therefore difficult to be reliably and consistently replicated. In addition any detailed analysis or changes in the representation or update of the data, are very difficult to achieve efficiently.

However, until recently GIS has only been used as an enhanced version of the original paper-based approach to HLA, resulting in simple digital maps. This may solve some of the problems of replication and representation of data, but these lack inherent intelligence, and importantly metadata, which are key aspects of true GIS applications. These issues are now being addressed within various county methodologies. However it is important to remember that GIS is not an end-use in itself, but a tool that facilitates input, representation, access, analysis and output of coherent data, to an established standard.

In response to these parameters, the methodology has been refined and developed in the light of other HLC projects, to accommodate regional diversity and harness the capabilities of GIS. This has been achieved, by creating a suite of 'historic landscape types' or 'attributes'. These may then be aggregated as required for different objectives, remits and end uses (Dyson-Bruce *et al.* 1999: Fairclough 1999). The attributes are derived either directly from map sources, *e.g.* woodlands, informal parklands, or indirectly by interpretation by informed judgement, for example, field types forming identifiable 'footprints' or distinct morphology, *e.g.* co-axial fields, parliamentary enclosure. However in reality, due to its complexity it may not be possible to define the

landscape into large morphologically consistent geographic units. The metho
flexible, so this suite of attributes may be expanded as required, in response to
diversity. In addition the inclusion of dated sources, directly within the databa
specific time horizons to be recognised.

HLA as a methodology has necessitated this fluid and dynamic approach 1
and record the subtleties and dynamics within the landscape. This has had to be
suitably sensitive manner, to ensure the development of a more rigorous anal
relating to academia, data entry, analysis and metadata.

8.4.3 Application

The revised HLA methodology, used in the East of England Project is now C
and may be described by the following three interactive aspects:

- **HLA** – creates 'historic landscape types', attributes that may be aggregate
  'historic character areas' – *the academic, archaeological and historic*
  These are recordable historical events within the landscape, derived from
  of sources.

- **GIS** – which handles the data capture process and input – *the practic*
  Ensuring consistent application and data management.

- **Metadata** – is the data informing the user of the process of, for exam
  collation, source, ownership and map creation, and is imbedded within the
  In addition it creates an audit trail of data synthesis, analysis and interpreta

8.4.4 Resources and Sources

Due to the constraints of staff, time and resources, the HLA methodology is by 1
quick, broad-brush and 'desk-top' based approach. It is meant to rapidly a
current landscape as to its historic origin and character. This is achieved by
primarily historical information.

Key sources include maps (current and historic, paper and digit:
photographs and documentary sources. The criteria used must be robust,
replicable and meaningful not only to the data creator, but to a variety of
objectives. However the landscape types must be sufficiently sensitive and de
reflect landscape composition, diversity, variability, continuity and discontinu
enables the complex concept of 'time-depth' and palimpsests within the lands
assessed.

In moving from one county to another the availability of digital/pape
varies, as does quality of that data, hardware, software and support and these c
direct impact on work progress and efficiency. For example, datasets may not
created at an appropriate scale, or may have used different criteria, or b
metadata, which at times renders potentially useful data unusable. In addi
datasets, which may have been created independently, usually at differing
difficult to combine due to among other things edge-matching problems. It is 1
with the migration to the new OS MasterMap this will be a problem of the past.

8.4.5 Data Collation

The data is initially collated at 1:25,000 (data analysis), as this represents the smallest scale at which field boundaries are represented on OS topographic maps (1:10,000 sheets were deemed too detailed and cumbersome). The 1950 1:25,000 OS series (paper maps), have been used to collate the first tranche of information, being compared to the current digital OS landline and 19th century First Edition OS data, giving three distinct time horizons.

In certain parishes additional map sources were used for research purposes to try and further understanding of the complexity of field morphology. Tithe, Enclosure, and early estate maps were consulted, to inform the process of field shapes, types and land enclosure, which has expanded the series of time horizons within these specific areas.

This information is then digitised, on screen, using a defined series of landscape types (attributes) to create complete coverage. In the future with the forthcoming, polygonised OS MasterMap, digitising will be much reduced, with data being seeded into the pre-existing polygons.

At the generic level these landscape types distinguish the various landscape components, for example field systems, woodlands, parklands, grazing, urban, industrial, extractive, military (both current and relict of each type). These attributes must still be currently visible within the landscape to be recorded. Recording the data in this way enables it to be used flexibly in response to a variety of objectives. For instance, this data may then be aggregated, as desired, to observe broad patterns in historic land-use at a county or regional level.

8.4.6 Metadata

The metadata has been incorporated directly within the database, with appropriate fields which includes:

- *historic landscape type* – these fields allow short succinct data entry via pre-set 'alpha-numeric' codes which are linked to another database providing full descriptors of landscape type (*e.g.* code_c, relict_1 or Hla_code). For example sf, represents sinuous fields; ip = informal parklands; ba = built-up area; aw = ancient woodland; pf = prairie field. These are based on direct or indirect evidence of landscape component type, either untampered through time or as a palimpsest with various events recorded within a composite unit. It follows that the more events recorded, the more fragmented the landscape. Conversely a single relatively untouched event may cover a much larger area, *e.g.* a block of parliamentary enclosure. So far up to four events have been recorded within a landscape unit, however this does not reflect an older historic origin, purely that more events or changes have been recorded within that unit of space.

- *data source* – the maps from which the data derived, whether current or relict, *e.g.* date_c, date_r1. These fields again provide a single character link to that landscape's component source, *e.g.* L = OS Landline 2001, N = First Edition OS. This creates a clear audit trail of source synthesis, as each polygon may reflect several sources in its creation. However it has been noted that in some instances there have been changes or events within the landscape which fall between these

time horizons, and are therefore not able to be adequately recorded, *e.g.* u
mineral extraction, then recorded land restoration.

- *creator* – this records the persons responsible for assessing, assigning
deriving the original polygon representing the landscape component.

- *date of creation* – date the data was digitised, which informs currency of u
data creation.

- *data owner* – this reflects which county or regional authority has owne
copyright of the resultant material, *e.g.* Essex County Council (ECC), Hei
County Council (HCC).

- *scale & consistency of data input* – the scale of digitising is set at
Whereas the source material is primarily at 1:25,000, it has been enha
checked at 1:10,000 with heads-up, on-screen digitising at that scale,
consistency.

- *glossary of terms* – *e.g.* code_c, code_r2. This is a linked database prov
textual descriptions of landscape types, so that users can understand the
types without having to use a lookup table to interpret the codes, whilst
progress.

8.4.7 Map Sources

In addition there are linked databases to correlate data, especially to the dat
primarily the maps used as a source to create the HLA dataset. It is important to
in assessing map data, especially historical sources that these may reflec
objectives and that paper maps can only represent limited and selected informa
raison d'être of creation may be different to its current use and interpretation, an
important caveat. For example some of the early county maps are more a ref
wealth than geographic reality, *i.e.* one paid to have ones estate/manor represent
map. Other maps are records of resources, for example the Roy maps of Scotla
records primarily resources of strategic use to the military, *e.g.* woodlands,
routes, water, populated areas.

    In addition there is a major problem of data inconsistency not only
different counties and regional authorities, but within, and this includes paper a
sources. This creates a practical problem in consistency and accuracy in data
With paper maps all georeferencing is visual, whereas digital maps are georefere
therefore may be easily overlaid for comparison and analysis. In reality, one
assessing an avalanche of paper maps, or a very cluttered GIS screen with many
you have both – usually the latter. There are benefits and disbenefits to having
either digitally or as map sheets, experience has proven that it is better to have b

    The datasets that are consistent between counties include the OS Landl
1:10,000K raster maps and the 19th Century First Edition (either paper or dig
1950 1:25,000 OS paper maps. The sources that are consistent within a count
the 18th Century county series, *e.g.* Dury 1760 for Hertfordshire or Andre &
1766 for Essex (all paper sources). In addition sources that are inconsiste
counties have been referenced and include parish based historical sources, *e.g.*
Enclosure maps, and individual Estate maps. These vary in size, scale (if any is

detail, accuracy and quality. Few have been turned into digital sources, as many vary in size, are fragile, unwieldy and unique historical documents, at times very large covering map tables requiring careful unrolling, one being able to see only small sections at a time. Up to date aerial photography has only recently been digitally available as a georeferenced source, otherwise they require visual referencing.

It is hoped there will be links within the database to other sources of information, which will include other counties methodologies, *e.g.* Suffolk: hyper-links to photographs, scanned images, maps, textual documents, videos. These databases in conjunction with the embedded metadata, will therefore help to render the methodology as transparent as possible and be self-documenting for future researchers and users. In addition the HLA data may be used as an archive, the data may be 'static', providing uniform time horizons or 'dynamic' allowing constant revision.

8.5 DISCUSSION

The East of England Project is unique, as other counties within England are being, or have been assessed, independently using a variety of methodologies. The experience of the project so far has shown that there are distinct advantages to the application of a single methodology to a group of counties being assessed as a single geographic region. This ensures a seamless map with consistency in methodology, application and analysis ensuring replicable results yielding compatible analysis. Therefore analysis should reflect historic landscape diversity rather than methodological differences.

Metadata is now being increasingly recognised as an important facet of GIS, to inform the user of the very nature of the data. It is often a misunderstood or poorly understood concept outside the realm of GIS practitioners. Its inclusion within the database of HLA ensures a clear audit trail of synthesis, scale, ownership etc.

Broad historic landscape patterns are beginning to be identified which appear in some instances to disregard county boundaries, an important consideration when one sees how an individual county's methodology and application respects each county's boundary. Some landscape types are scattered through the countryside, others have a geographically limited spatial distribution, or are mutually exclusive. However, there may be localised pockets of distinctive historic landscape character and development. This patterning reflects landscape diversity in historic terms. It is proving to be a diverse, complex and dynamic landscape (see Figure 8.2).

In collating the data the inconsistency and availability of datasets, GIS software, hardware, resources, GIS support all have direct impacts on an application such as this. It is therefore important to recognise this and take such realities into account when devising a methodology which is not only meant to be a simple, direct, practical GIS application, but also a pragmatic management tool, and is in addition an academic exercise.

When specific areas have been completed, with geospatial analysis and additional research it is hoped to identify why some areas exhibit these differences in diversity and contiguity of landscape character. Initial research indicates a range of factors including, for example, historic tradition, social, economic, political, agri-environmental, geology, soils, slope, aspect, and height. These all have a part to play in landscape development. It has yet to be established in what proportion, why, and how these various factors influence the historic development of the landscape, and could form the platform of future research.

However any GIS research will be limited to those datasets which are available at an appropriate scale, which again may have disparate regional availability.

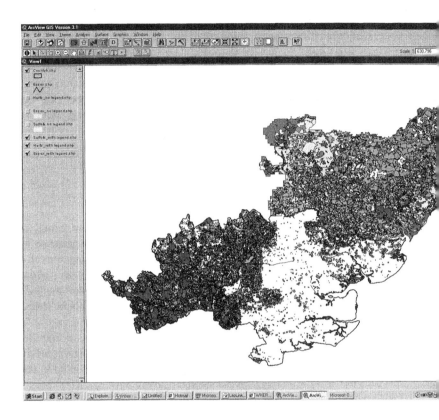

Figure 8.2 East of England Historic Landscape Assessment: Suffolk, Hertfordshire and Essex - work in progress.

HLA has already proven to be a valuable management tool, for example i and informing:

- Local, regional, historic landscape patterns
- Chronological patterns, through time and space: landscape 'time palimpsests
- Landscape management and Landscape Character Assessment
- Management of the cultural resource
- Archaeological survey
- Research objectives
- Development control in planning
- Regional and Local Planning issues, policies and strategies

However the author also feels it is becoming increasingly important that GIS strategy be implemented to ensure a consistency in standards and applicati within this field of historic time-series analysis. In addition a consistent methodology and use of terms would encourage use and respect of such a dataset, especially in controversial and contentious circumstances such as Public Inquiries. Planners, landscape architects, developers and other users could have faith in the data for improved land management, research and other applications.

8.6 CONCLUSIONS

The HLA methodology has necessitated a flexible and dynamic approach to respond and record the subtleties within the landscape in an appropriate manner. In addition the methodology has been required to take into account GIS issues relating to metadata, including data source, synthesis, entry, and analysis.

Experience is proving that appropriate hardware, software and a robust methodology are essential for the success of any GIS project. It requires clarity of thought and application, for a GIS map/theme/layer/coverage is only as intelligent as the database and metadata supporting it. Realistic objectives and time-scales with appropriate resources need to be established and made available, to enable the successful completion of any project.

The change of methodology from that of 'paper-based' to GIS approach, has not been without difficulties, but is constantly under improvement and being refined with experience and expertise. This is leading to a more robust, transparent, and flexible methodology. This will facilitate a more reliable analysis not only within, but also between, different counties. Thus results obtained will reflect real and meaningful historic patterns within the landscape, rather than differences of methodology in the process of assessment, application, synthesis and analysis.

Therefore the incorporation of metadata within the database is fundamentally important in ensuring a transparent methodology to help maintain consistency of application and analysis between regions, by a variety of practitioners. This particular application focuses on specific time-horizons, incorporated within the database to widen the scope of current and future synthesis and analysis.

Finally it is proposed that a national GIS set of standards and Historic Landscape Types be formulated to enable appropriate and consistent GIS methodologies be established to ensure compatibility of results. Thus local, national and regional landscape models may be developed, which will consistently reflect true landscape diversity across England. There is no doubt that in the future, the analysis and management of this form of historic spatial data lies within the GIS platform.

8.7 ACKNOWLEDGEMENTS

The project has been funded by English Heritage, and supported by the relevant local authorities.

I should like to thank Graham Fairclough (English Heritage), Stewart Bryant (Hertfordshire County Council), Paul Gilman (Essex County Council) and Edward Martin (Suffolk County Council) for their assistance and support in this project.

(The views expressed are those of the author and are no reflection of English Heritage (EH) or the associated local authorities).

8.8 REFERENCES

CADW, 1998, *Register of Landscapes of Outstanding Historic Interest* (Cardiff: CADW).

Countryside Commission, 1997, *Landscape Assessment Guidance.* C Commission.

Dyson-Bruce, L., 1998, *Historic Landuse Assessment*, Unpublished R Historic Scotland.

Dyson-Bruce, L., with Dixon, P, Hingley, R., Stevenson J., 1999, *Historic Historic Landuse Patterns. Report of the pilot project 1996-98*, Researc (Edinburgh: Historic Scotland & RCAHMS).

Fairclough, G., (editor), 1999, *Historic Landscape Characterisation,* (London Heritage).

Ford, M., 1999, *Historic Landscape Characterisation in Suffolk 1998-199* Report.

Herring, P., 1998, *Cornwall's Historic Landscape: Presenting a Method o Landscape Character Assessment.* (Truro: Cornwall Archaeology Unit).

Wrathmell, S., Roberts, B.K., 2001, *Historic Settlement Atlas of England* Heritage.

9

Using GIS to research low and changing demand for housing

Peter Lee and Brendan Nevin

9.1 INTRODUCTION

This paper explores the sustainability and popularity of neighbourhoods using administrative data at neighbourhood, local authority and regional level in the UK. The paper is based on a number of research projects conducted by the authors on the issue of low and changing demand for housing in the North West and West Midlands conurbation's of the UK. The research is principally dependent on the use of GIS in determining coterminous boundaries for the analysis of social cohesion, social exclusion, access to community facilities and housing popularity. The research identifies a number of problems associated with the identification of unpopular or low demand housing areas based on the analysis of poverty at small area level and illustrates the value of GIS in assisting with the targeting of resources and the formulation of local urban and housing policy.

During the period 1992-2001 the UK experienced the longest duration of unbroken economic growth since 1945. However, whilst regional disparities in unemployment rates fell, highly significant differences in regional and sub-regional housing markets emerged. A shift in housing demand away from the north to the south of England was partly explained by differential migration patterns (Holmans and Simpson, 1999). This lead to social housing providers in some localities attempting to manage significant increases in empty properties and turnover in the social rented sector. But whilst, publicity surrounding this issue initially focused on the difficulties being experienced by the social rented sector, it became apparent that parts of the private sector owner-occupied market were also experiencing problems evidenced by low sales prices and in extreme cases abandonment of properties. The spatial manifestation of these housing market changes has increasingly been focused on disadvantaged neighbourhoods and in some areas in the North and Midlands, this has resulted in the progressive abandonment of neighbourhoods. The publication of a number of UK government and academic reports since 1999 (see DETR, 1999, 2000; Lowe *et al.*, 1998) succeeded in raising the issue of low and changing demand within the policy community. However for housing professionals the literature at this stage did not greatly assist them with targeting resources at neighbourhoods or housing markets in decline, particularly in districts where deprivation was widespread.

In this paper we want to demonstrate how the issue of low demand[1] was r
in different local and regional contexts using GIS and spatial analytical techni
paper makes the point that it is important to differentiate between
neighbourhoods that are sustainable and those that are deprived and app
dysfunctional. GIS can help in differentiating between these spatial trends. We
that the issue of low demand is well suited to spatial analysis as the dynamics c
changing demand is influenced by a number of processes taking place at differ
scales, which cut across local administrative and political boundaries. The
draws on three separate case studies conducted by the authors at local and su
level. The creation of new policy boundaries cutting across pre-existing ma
administrative/political boundaries is an attractive feature of GIS tools, and
policymakers find useful in clarifying issues or highlighting new strategic priori

9.2   CAUSES OF UNPOPULAR HOUSING AND CHANGING DEMANE

Explanations for low and changing demand differ from area to area. Not onl
specific analyses important but the level of analysis is also important in unde
different process occurring at regional, local and estate level. It is undoubtedl
that at the micro-level there are a number of factors that may explain pop
neighbourhoods and housing satisfaction and which in turn affects housing dem
influence of 'problem' families on an estate or the effect that ma
intervention/non-intervention has upon the popularity of neighbourhoods are im
this respect. Socio-psychological explanations (Bruin and Cook, 1997), prefe
different types of neighbourhoods according to resident perceptions (Adams, 1
role of successful role models (Wilson, 1989) and ethnic segregation (Galster, 1
also been explanatory factors used. Multivariate approaches have been used t
household surveys in order to explain resident satisfaction (Canter and Re
Ginsberg and Churchman, 1984) but fail to incorporate an analysis of broade
market dynamics. In this respect, Lu (1998), shows that structural factors hav
direct effect on housing mobility intentions independent of resident satisfaction.

In this paper we wish to focus on broader spatial patterns of neighbourho
and how these relate to other contextual factors such as poverty. Murie *et*
identify the most significant macro level factors influencing whether low d
present and these include:

- The quality and condition of the housing stock;
- Residualisation of social housing due to increased identification of the se
  lower income groups;
- Reputation of the area;
- Aspirations of new entrants to the housing market;
- Changing markets resulting from a private sector building industry able t
  to the changing aspirations of consumers in an improved economic en
  where choice is important;
- The age profile of parts of the housing market especially where it is heavi
  towards the younger and older age groups.

---

[1] We use the terms 'low demand' and 'unpopular housing' in an interchangeable way but differenti
low demand (housing in excess of those prepared to rent) and 'changing demand' (a movement
particular tenures or dwelling types).

9.3 RESEARCH ON UNPOPULAR HOUSING AND LOW DEMAND

9.3.1 Measuring Sustainable Housing Areas in Liverpool

We begin with research in Liverpool conducted between 1999-2001, which has contributed to the city's housing strategy statement and guided decisions on unsustainable and unpopular housing areas. Here we describe the way in which GIS was used to explore the manifestation of problems at the neighbourhood level.

The housing market in Liverpool is highly differentiated by price, quality and residential turnover. The city has been vulnerable to market change as it has a larger than average social rented sector and a disproportionate number of low value terraced housing renovated during the 1970s and 1980s. many of these properties are now nearing the end of their extended life and the combination of these housing factors as well as long-term economic decline and population loss meant that Liverpool has experienced both changing demand and low demand for housing across all tenures. In 1999 the vacancy rates for all stock in the city was more than double the national average. It was against this background that Liverpool City Council and the principal Registered Social Landlords (RSLs) in the city commissioned the Centre for Urban and Studies (CURS) to examine the issues surrounding the changing demand for housing and the sustainability of neighbourhoods in the city.

We began by differentiating between contextual indicators of unpopularity and direct outcome measures of housing and neighbourhood unpopularity. We developed 4 separate domains of sustainability with which to measure neighbourhood sustainability and housing unpopularity at the small area level:

9.3.1.1 Poverty and Social Exclusion

As poverty is widespread in Liverpool and as previous research in Birmingham had shown us that not all deprived areas are unpopular (Lee, 1998), however, it tends to be a precondition for circumstances leading to low and changing demand. Four poverty outcome measures were used:

- *Council Tax and Housing Benefit claimants*: claimants in receipt of housing benefit identified from the city's Council Tax register and aggregated to enumeration district (ED) from postcoded information;

- *Children in receipt of free school meals*: Local Education Authority supplied data on children in receipt of meals aggregated to postcode and matched to ED;

- *Unemployment*: unemployment aggregated from postcoded information to ED;

- *Standardised Mortality Ratios*: aggregated to ED from health authority records on deaths.

9.3.1.2 Crime and Social Cohesion

Poor social networks at the neighbourhood level and its impact on social cohesion and crime (Freudenburg, 1986; Perry 6, 1997; Gordon and Pantazis, 1997) as well as the known relationship between crime, poverty and social cohesion (see: Hirschfield and Bowers, 1997) lead us to develop a measure of social cohesion based on the following measures:

- *Property Crime and Crimes Against the Person*: the different geog▮ sociology of crimes lead us to separate out property crimes and crimes a▮ person. Merseyside Police supplied data at ED level. Since 1999 it ha▮ easier to obtain data on crime at small area level due to legislation pror▮ community safety role of local authorities.

- *Population Decline*: Using 1991 census as a baseline, 1997 mid-year ▮ estimates at ED level were used to estimate change. Areas with a ▮ population used as a proxy for instability and unpopularity;

- *Voting Turnout*: ward level figures on voting turnout at elections i▮ supplied by the local authority and used as a proxy for social cohesion.

9.3.1.3 Environment and Infrastructure

Environmental infrastructure and amenities and their proximity or accessi▮ determine quality of life. Burrows and Rhodes (1998) found that the t▮ important factor in determining neighbourhood satisfaction is the presence or ▮ leisure facilities. This domain therefore attempts to measure this dim▮ popularity.

- *Percentage of Derelict Land within an ED*: the planning department ▮ polygon file of known derelict land sites. This was over laid on to an E▮ file and the proportion of the surface area currently derelict calculated;

- *Number of Amenities*: the planing department also supplied detailed infor▮ the location and type of non-residential addresses. These were match▮ object files in MapInfo using SQL queries to select out different types of a▮

9.3.1.4 Housing Unpopularity

The housing unpopularity domain was used as a direct measure of low de▮ unpopularity and treated as a dependent variable incorporating three measures: ▮ voids and number of void days and house prices.

- *Vacancy Quotient 1993/99*: A vacancy quotient is a measure of both the▮ time that a property is void and the proportion of properties that become▮ given period (Smith and Merrett, 1988). The Council Tax register for ▮ April 1993 to April 1999 was used to identify periods over which prop▮ become vacant as well as vacant properties as at April 1999.   Prope▮ matched to ED using AddressPoint in MapInfo.

- *Average House Prices*: House prices aggregated to Postcode Sector by ▮ obtained from HM Land Registry for the period 1995-1997. EDs wer▮ house price values by matching ED centroids to postcode sector ce▮ MapInfo.

All the data was standardised around a mean of 0 (Z Score) in SPSS before ▮ were transferred to MapInfo. Using MapInfo we generated choropleth maps ▮ separate domains as well as observing the statistical overlap between then▮ poverty is widespread across the city of Liverpool, Figure 9.1 shows tha▮ unpopularity is most problematic in the Inner Core of the city as characteris▮

house prices and high vacancy rates and episodes. Figure 9.2 shows the coincidence of areas with high scores on the housing unpopularity domain that also have low social cohesion and high rates of poverty are mostly located in the Inner Core of Liverpool.

Using GIS to differentiate between elements of housing market weakness at small area level and to highlight the spatial coincidence of area deprivation and unpopular housing areas allowed several policy recommendations to be made (see Nevin *et al.*, 1999). The City Council have adopted the recommendations within the report and the Annual Housing Strategy Statement for the years 1999-2001 have focused investment and resources as the different housing issues prevalent within the housing zones highlighted in Figure 9.2.

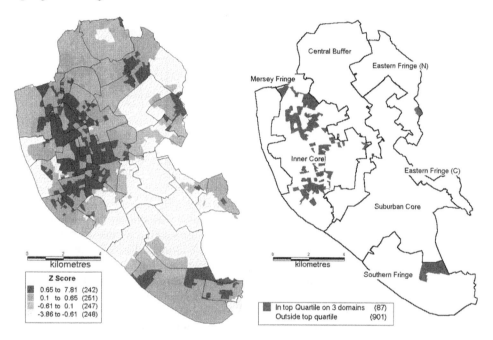

Figure 9.1 Unpopular housing areas in Liverpool.

Figure 9.2 Liverpool policy zones: multiple problems associated with housing market failure.

9.3.2 The Relationship Between Poverty or Social Exclusion and Low Demand for Housing

The policy debate on low and changing demand and unpopular housing within the UK has largely been orchestrated from within the Cabinet Office's Social Exclusion Unit (SEU). The SEU's Policy Action Team report on Unpopular Housing (DETR, 1999) puts a heavy emphasis on the role that poverty plays in causing low demand (DETR, 1999, p.25). But, how far does poverty, using the Liverpool case study, explain variations in housing unpopularity, or relate to the other known drivers of unpopularity and low demand?

Table 9.1 shows the correlation coefficients between the housing unpopularity domain and the other domains. Whilst all the coefficients are positive the overlap

between the individual domains is far from the maximum. This indicates that s
areas are socially cohesive with lower than average crime rates and population
policy implications would indicate a different set of responses in 'low crin
poverty' areas than in areas where crime and poverty are equally problem:
standardised values of the individual domain indicators were used in a
regression model to explain the variation in levels of housing unpopularity in v
domain was used as the dependent variable. Across the city 56% (R=0.74(
difference in levels of unpopularity was accounted for by the standardise
indicators. While this is a reasonably large amount of explained variation ther
44% unaccounted variation in levels of popularity.

Table 9.1 Correlation Coefficients for the Four Domains.

	Poverty & Social Exclusion	Crime & Social Cohesion	Environment & Infrastructure	H< Unp<
Poverty & Social Exclusion	1.00	-	-	
Crime & Social Cohesion	0.64	1.00	-	
Environment & Infrastructure	0.16	0.04	1.00	
Housing Unpopularity	0.70	0.61	0.06	

The unexplained variation will be accounted for by variables not measure
in the analysis (examples may include attractiveness of housing on offer, labo
role of the area, number of students living in the area and so forth). Therefore,
would not argue against the idea that poverty is a precondition for low demai
analysis using GIS at neighbourhood level in Liverpool shows, we would empl
not all poor areas are unpopular or suffer from low demand. It is therefore im
differentiate between observable measures of poverty and the dynamics o
market change which can lead to unpopularity and area blight. Additionally
these processes coalesce is important. Areas with observable scores that are
adjacent to relatively stable and popular neighbourhoods will have a different
and require a different set of responses compared to areas that score highl:
surrounded by similar areas.

9.3.2.1 Adjacency Analysis

GIS is a useful tool in this regard as it allows us to develop a method by whi<
compare adjacent small areas (Enumeration Districts or EDs) and identify coak
problems of unpopularity or areas at risk of low demand. In order to compan
area scores we created two identical object files with attached data fields and ir
ran an SQL command intersecting all objects on the two files (Figure 9.3). The
query file (subsequently saved as *ed_y_h_pairs*) is all the possible pairs of adj:
with the original data value (*index_p*) and its paired values (*index_p_2*) (Figure

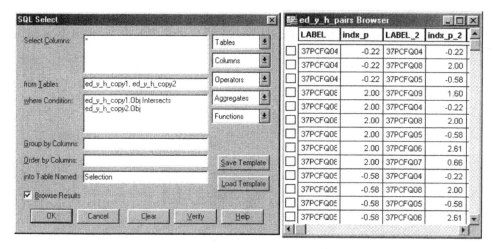

Figure 9.3 SQL Intersect command in MapInfo on two identical object files.

Figure 9.4 Resulting 'paired' objects with associated data.

In the next stage of analysis we aggregated the values of all the adjacent EDs using an SQL command in MapInfo (Figure 9.5). The resulting values are shown as the original *index_p* value for each object as well as the *average* score based on all adjacent pairings of EDs (Figure 9.6).

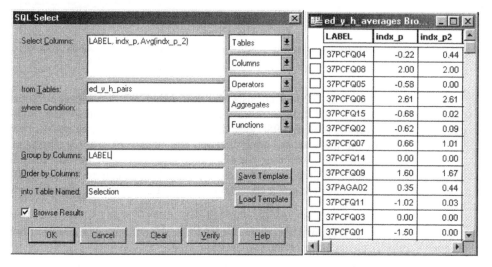

Figure 9.5 SQL command in MapInfo aggregating adjacent object data values.

Figure 9.6 Resulting 'aggregated' data file and object label.

9.3.3 Measuring Unpopularity in the North West of England

Using the same 'adjacency' methodology we showed that more than 95% of EDs within the North West M62 corridor (8555 of 8987 EDs) are adjacent to or scored above expected on the government's area index of deprivation (the ILD – Index of Local

Deprivation; DETR, 1998) (Figure 9.7). The implications confirm that d
appears to be widely spread across the North West and that deprivation i
instrument in targeting areas at risk of low demand.

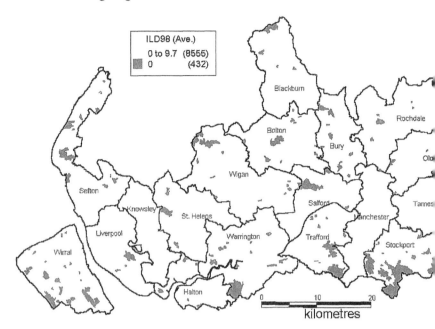

Figure 9.7 EDs not adjacent to deprived areas: average scores based on adjacency pairings (M62
Source: Nevin *et al.* (2001a).

The M62 Corridor consists of a consortium of 18 local authorities (see Figure
experienced a 22% increase in local authority vacancies and a 72% rise in F
over the period 1995-1999. The authors were commissioned to identify the co
of the most serious problems and the areas most at risk of low or changing
Drawing on previous research (including the Liverpool study) we therefore id
number of factors influencing changing demand and helped us focus on the eler
we believe are important in determining areas at risk of changing demand. The
included:

- areas or parts of the city in which there is a predominance of rented housii
  quality stock in owner occupation;

- neighbourhoods in which there is a large-scale or monolithic pro
  'obsolescent housing' of a certain type, *i.e.* high rise flats or terraces;

- areas with demographic characteristics likely to weaken demand such
  concentration of elderly residents;

- a concentration of households that are economically inactive or unemploye

An index was produced which mapped the areas at risk of changing or lov
in the sub-region. The index was constructed in two parts:

1. social housing areas (predominant housing is public sector) with higher than expected combined standardised values on over 65s, economically inactive, unemployed and flatted accommodation;

2. private housing areas (predominant housing is private) with higher than expected combined standardised values on over 65s, economically inactive, unemployed and terraced accommodation.

Indicators in both indexes were standardised using chi-square in SPSS and combined into one single index by substituting the highest value in each case. The index values were adjusted in MapInfo by taking into account adjacent ED values as in the methodology described above.

This method enabled us to identify coalescence of problems highlighting in particular clear spatial concentrations of neighbourhoods at risk of changing demand in the core of the Greater Manchester conurbation (Manchester, Salford and to a lesser extent Trafford) and in the inner core of Merseyside centred on the City of Liverpool, Bootle, Tranmere and Birkenhead. Smaller areas of potential decline are also highlighted in St Helens, Halton, Wigan, Warrington, Blackburn with Darwen, Bolton, Oldham and Rochdale (see Figure 9.8). There are 280,000 households contained within the overall clusters of areas at risk of changing demand (16.3% of the households in the study area). These areas contain a population of 690,000 people (15.9% of the population of the M62 Corridor). Neighbourhoods at risk are predominantly social housing areas, however there is clear evidence of multi-tenure problems with nearly 100,000 properties being privately owned. These multi-tenure issues are most pronounced in the Merseyside Inner Core where 46% of households either rent privately or own their homes.

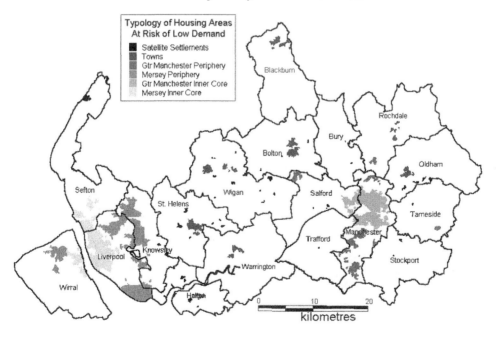

Figure 9.8 Typology of areas at risk of changing demand in the North West M62 Corridor.
Source: Nevin *et al.* (2001a).

The research maps of areas at risk were produced to enable the identif individual enumeration districts and were distributed to each of the agenc Commissioning Consortium. The responses received indicated that the Risk identified neighbourhoods where problems of low demand were entrenched a where problems were beginning to appear. The analysis is now being used housing investment resources and has been given significant emphasis in Regional Housing Statement for the North West of England.

9.3.4 Validating the Risk Index: A Case Study of Birmingham

The methodology was subsequently employed in a study of changing housing r the West Midlands Region (Nevin *et al.*, 2001b). The region has a population 5 million in which Birmingham is the most densely populated city, housing mc million residents. As at June 2001 Birmingham City Council owned 88,000 and was pursuing a policy of wholesale stock transfer to eleven new Register Landlords. In October 2000 the city commissioned the University of Birmi assess the popularity of housing in order to make decisions concerning the trans city's housing stock.

To test the accuracy of the Risk Index for Birmingham (Figure 9.9) w together two additional proxy indicators of 'unpopularity':

1.   lettings of local authority stock between 1990-2000 (Figure 9.10);
2.   transfer requests from local authority stock (as at October 2000) (Figure 9.

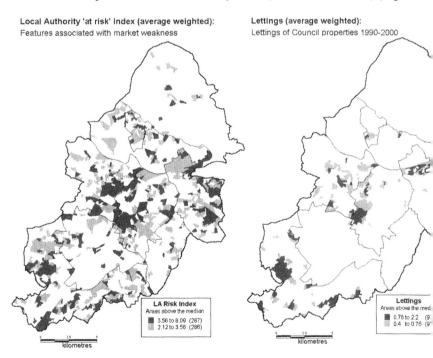

Figure 9.9 Local Authority Areas   **Figure 9.10** Letting of council prope
'at risk' of Low Demand.   (standardised, weighted average), 199

These were used to highlight local authority housing areas with market weakness measured on all three dimensions of 'unpopularity' in which areas scored highly on all three measures. In this study negative indicator scores were set to 0 in MapInfo before average indicator scores were calculated. The scores were re-set to 0 in cases where the original score was zero. As the study was primarily concerned with the popularity of local authority stock we also set to zero scores in areas where there was no local authority stock as at 1 October 2000. Areas were assigned a score of 1 if they scored in the upper median on any one factor: 'at risk' of changing demand, higher than expected lets or higher than expected registered transfers, the breakdown for the city as whole is as follows:

- Of the 1,485 enumeration districts containing council housing, slight more than half featured in the bottom median on all three measures;

- A further 32% (471 EDs) had a score above the median on at least one measure;

- The remaining 258 EDs (17.4% of the total containing council housing), were split as follows: 187 EDs (12.6%) had scores above the median on two measures whilst slightly under 5% of EDs (71) had scores above the median on all three measures.

The results of this exercise have shown a high correlation between those areas which we would have expected to show signs of decline given their housing and socio-economic characteristics in 1991 (using the 'at risk' index for the North West and West Midlands) and what has actually happened over the period 1990-2000. For example, in most neighbourhoods highlighted in Figure 9.12, neighbourhoods designated at risk of low demand have either high turnover or transfers or both.

Registered Current Transfer Requests (average weighted):   Areas showing signs of unpopularity
Transfers registered after 1 January 1990   EDs scoring above median on 2 or more measures

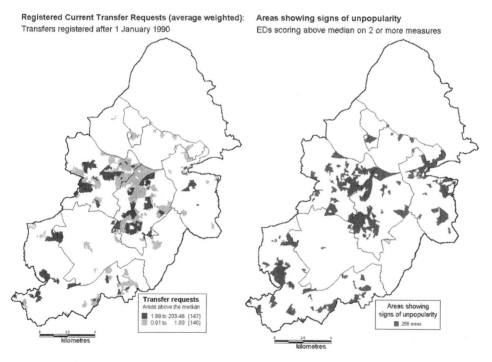

Figure 9.11 Registered transfers   **Figure 9.12** Unpopular council housing areas.
(standardised, weighted average), 1990-2000.

However, there is not a perfect fit and in a number of areas the problems o
and transfer requests are more widespread than the Risk Index would predict.
point illustrates the need to check data prepared using GIS against other
measures such as crime, environment and property construction. These find
been developed further to assess the drivers of change in different areas of the
have been translated into policy responses in the City's housing strategy.

9.4   CONCLUSIONS

This paper has illustrated the use of GIS packages such as MapInfo in
administrative data for the purpose of categorising areas with characte
unpopularity or experiencing problems related to housing demand. It is clear th
considerable difficulty in determining the viability and sustainability of hou
both empirically and methodologically. Whilst the focus of this paper has be
empirical and technical aspects related to this issue, the task cannot simply be r
a technical exercise alone in which areas are ranked and reduced to an index so
agreed scale. The policy agenda and the management practices of social ar
landlords as well as broader regional and sub-regional trends have to be incorpo
the analysis. It has not possible for us to explore these themes here. However,
argue that the use of these techniques and the differentiation between pove
cohesion, housing popularity areas at risk of changing demand needs to be th
point for exploring policy solutions. In this way it enables policy makers to ta
for investment or disinvestment on the basis of market and socio-
characteristics. This is particularly valuable in local authorities where depr
widespread. Additionally the mapping of areas at risk using GIS has proved to
effective in conveying complicated messages to a varied audience which has
politicians and tenants and residents groups.

There are a number of implications for further research in this area and
other administrative data and national data sets. What this paper sets out to sh
contextual indicators such as poverty, although widely referred to in the literatu
discriminate sufficiently between popular and unpopular areas. For this reas
direct measures of low demand should be used. Moreover, where regener
housing strategies have failed in the past is partly due to the focus on static ge
and narrowly drawn target areas. In this paper we have illustrated ways in
coalescence of problems associated with low demand can be highlighted thi
application of 'adjacency' analysis. It would not be possible to conduct such
without the aid of a GIS or mapping package. Additionally, we have attempted
the problem of static geographies by conducting regional and sub-regional analy
we feel is crucial to the study of low demand as it is the different geographical
enquiry that help us understand the dynamics of low and changing demand a
new policy vehicles to be deployed which cut across existing administrative bc
Whilst it is costly and time consuming to assemble data for a large number of a
modelling census data has enabled us to construct a picture of risk across sub-
small area level. By mapping direct indicators of low demand from admi
records at local authority level in Birmingham and Liverpool we were able t
this analysis. It will therefore be possible to conduct a similar exercise using
census to identify areas at risk of changing demand over the next decade.

9.5 REFERENCES

Adams, R.E., 1992, Is happiness a home in the suburbs - the influence of urban versus suburban neighborhoods on psychological health. *Journal Of Community Psychology*, **20**(4), pp. 353-372.

Bruin, M.J. and Cook, C.C., 1997, Understanding constraints and residential satisfaction among low-income single-parent families. *Environment and Behavior*, **29**(4), pp. 532-553.

Burrows, R. and Rhodes, D., 1998, *Unpopular Places? Area Disadvantage and the Geography of Misery in England.* (Bristol: The Policy Press in association with the Joseph Rowntree Foundation).

Canter, D. and Rees, K.A., 1982, Multivariate model of housing satisfaction, *International Review of Applied Psychology*, **31**(2), pp. 185-208.

DETR, 1998, *The 1998 Index of Local Deprivation: A Summary of Results.* Department of the Environment, Transport and the Regions. (London: Stationery Office).

DETR, 1999, *Report by the Unpopular Housing Action Team*, Department of the Environment, Transport and the Regions. (London: Stationery Office).

DETR, 2000, *Low Demand Housing and Unpopular Neighbourhoods.* Department of the Environment, Transport and the Regions. (London: Stationery Office).

Freudenburg, W.R., 1986, The density of acquaintanceship - an overlooked variable in community research. *American Journal of Sociology*, **92**(1), pp. 27-63.

Galster, G.C., 1990, White flight from racially integrated neighborhoods in the 1970s - the Cleveland experience. *Urban Studies*, **27**(3), pp. 385-399.

Ginsberg, Y. and Churchman, A., 1984, Housing satisfaction and intention to move - their explanatory variables. *Socio-Economic Planning Sciences*, **18**(6), pp. 425-431.

Gordon, D. and Pantazis, C. (editors), 1997, *Breadline Britain in the 1990s.* (Aldershot: Ashgate).

Hirschfield, A. and Bowers, K.J., 1997, The effect of social cohesion on levels of recorded crime in disadvantaged areas. *Urban Studies*, **34**(8), pp. 1275-1295.

Holmans, A. and Simpson, M., 1999, *Low Demand: Separating Fact from Fiction.* (Coventry: Chartered Institute of Housing for the Joseph Rowntree Foundation).

Lee, P., 1998, *Popular and Unpopular Housing Areas: A Report to Urban Renewal.* Birmingham City Council.

Lee, P. and Murie, A., 1997, *Poverty, Housing Tenure and Social Exclusion.* (Bristol: The University of Bristol (The Policy Press) in association with the Joseph Rowntree Foundation, York).

Lowe, S., Spencer, S. and Keenan, P. (editors), 1998, *Housing Abandonment in Britain.* (York: University of York Centre for Housing Policy).

Lu, M., 1998, Analyzing migration decision making: relationships between residential satisfaction, mobility intentions, and moving behavior. *Environment and Planning A*, **30**(8), pp. 1473-1495.

Murie, A., Nevin, B. and Leather, P., 1998, *Changing Demand and Unpopular Housing.* Working Paper 4, (London: Housing Corporation).

Nevin, B., Lee, P., Phillimore, J., Burfitt, A. and Goodson, L., 1999, *Measuring the Sustainability of Neighbourhoods in Liverpool.* Research report by the Centre for Urban and Regional Studies, The University of Birmingham, for Liverpool First Partnership Group/Liverpool City Council.

Nevin, B., Lee, P., Goodson, L., Murie, A. and Phillimore, P., 2001a, *Changing Markets and Urban Regeneration in the M62 Corridor*. (Birmingham: C University of Birmingham).

Nevin, B., Lee, P., Murie, A., Goodson, L. and Phillimore, P., 2001b, *The West Housing Markets: Changing Demand, Decentralisation and Urban Reg* (Birmingham: CURS, The University of Birmingham).

Perry 6, 1997, *Escaping Poverty: From Safety Nets to Networks of Opportunity.* Demos).

Smith, R. and Merrett, S., 1988, *The Challenge of Empty Housing: Void Con Public Sector*. Occasional Paper No 31. (Bristol: SAUS, The University of B

Wilson, W.J., 1987, *The Truly Disadvantaged: The Inner City, the Underclass a Policy*. (Chicago: University of Chicago Press).

PART III

GIS and Urban Applications

10

Geographical visual information systems (GVIS) to support urban regeneration: design issues

Xiaonan Zhang, Nigel M. Trodd and Andy Hamilton

10.1 INTRODUCTION

Urban planning is complex. It combines data and opinion on social, political, economic and environmental issues. In western societies, planners are responding to new challenges to make planning more transparent, *e.g.* e-democracy, open government, and it has been argued that any solution should increase participation in the process. New technologies could be employed to promote dialogue and make the process less confrontational and easier to understand. The research at the University of Salford is designing and building a prototype which links emerging virtual reality (VR) and Internet technologies with (more mature) geographical information technologies. These geographical visual information systems (GVIS) are being designed to facilitate participation by multiple stakeholders in urban regeneration projects. In this chapter we consider how such a prototype can adopt a framework based on theories of learning and how that framework can be used to evaluate impacts of GVIS in urban planning.

The chapter introduces public participation in urban planning and reviews developments in GIS, VR and Internet technologies. A learning theory is outlined based on Kelly's theory of personal constructions (1955) and is mapped to functions available in GIS, VR and Internet technologies. Research is based on a case study of the Chapel Street Regeneration Project, Salford. The study identifies two stages of planning process – developing and selecting planning options – that offer opportunities for these technologies to engage stakeholders, and in particular the public, with the planning process. Design options for a GVIS are considered and methods of evaluating a learning system approach to GVIS are presented.

10.2 URBAN PLANNING AND PUBLIC PARTICIPATION

If a "city is a drama in time" (Geddes, 1905, in Cowan, 1998), then it is im
involve every citizen in writing the scenario and playing active roles in the d
sense of involvement not only gives citizens a greater meaning to their lives but a
a sense of responsibility which is often lacking in modern society (Ingram, 199
opinions are reflected in a movement from 'planning for the public' to 'plannin
public' (Klosterman, 1999; Roberts and Lloyd, 1999).

Whittick, (1974) defined public participation as 'the means by which m
the community are able to take part in the shaping of policies and plans that
the environment in which they live'. Participation should involve the publ
forward ideas and comments at the early stages of planning process (Sarjakoski
should be a continuous dialogue between the public and other stakeholders (M
1981). Rydin (1999) argues that in most planning activities participation is st
and low-level. It is usually undertaken in the late stages in the planning pr
mostly based on consultation documents and public meetings (DETR, 1998; B
and Walker, 2001). There is an argument for the public to be involved at earl
(Alterman *et al.*, 1984).

Two key factors have been identified for effective public participation:

- *Availability of, and Access to, Information*
  To participate in planning process, public require access to the necessary in
  about the planning work and have opportunities to express their opinions
  information about planning activities and the limited opportunities for partic
  planning policy decisions have been highlighted as key problems (Barlo
  Many authors propose that GIS technology could be used to increase public
  information and optimise their participation in the planning and polic
  process (*e.g.* Weiner *et al.*, 1995, Myers *et al.*, 1995 and Nedovic-Budic, 20
  envisage that people should have access to information presented at a
  understand and through media with which they are familiar.

- *Communication and Interaction*
  Planning requires a dialogue between and among stakeholders. Dialogue
  public and professionals can be inhibited by the lack of a common termin
  the alien jargon and alienating media used to convey ideas. Three-D repre
  of cityscapes offer more natural fora to exchange ideas (Sarjakoski, 1998
  Kodmany (1999) has shown that visualisation tools allow residents t
  participate in the design of their neighbourhood.

Nowadays, tools such as GIS, VR and Internet could be employed to ac
goals of effective participation.

10.3 DESIGN ISSUES 1: TECHNOLOGY

The last two decades have seen the dramatic development of information tec
and their ubiquitous adoption. In this section, the current use of three in
technologies, namely GIS, VR and Internet, in urban planning are outlined
integration of these technologies is considered.

10.3.1 GIS

GIS is a special information system, as the data it handles are all referenced with geographical location. It focuses on spatial entities and relationships and pays specific attention to spatial analytical and modelling operations (Maguire, 1991). GIS are a powerful tool for storing and handling geo-spatial data and have been adopted in many market sectors, such as telecommunications and natural resource management.

GIS research has, since the early years, "moved from primitive algorithms and data structures to the much more complex problems of database design, and the issues surrounding the use of GIS technology in real applications" (Goodchild, 1992). Although GIS are adopted widely, their potential remains unfulfilled (Batty, 1993; Douven *et al.*, 1993).

Until recently GIS users have been limited to specialists and professional users. This is attributed to two reasons:

- *Low Accessibility*
  Expensive software and data, poorly catalogued and protected databases are barriers to non-profit organisations and the general public (Nedovic-Budic, 1998). As a user group the public requires tailored small and beautiful GIS by which they can solve some simple spatial problems like *where is...?*, *what is at location...?* and *what if...?* by themselves.

- *Weak Visualization*
  The user interface is crucially important as it is the only part directly seen and 'is' the system for the user (Frank & Mark, 1991). Most commercial GIS-user interfaces are based on the use of windows, icons, menus, and pointing devices (Egenhofer & Kuhn, 1999). These interfaces are too often an impediment to effective problem solving or decision-making (Medyckyj-Scott & Hearnshaw, 1993).

The emergence of web-based GIS increases access. Through the Internet, people can transmit data and access tools to conduct analysis and create GIS presentations (Peng, 1997). Although the Internet-based GIS creates many benefits for the public, such as the convenience of access and the low cost, problems of tedious interface and difficulty of use are still not solved. Virtual reality (VR) offer potential to facilitate public use of GIS tools as it increases the engagement of the user by coming closer to natural ways of interacting with the world than would happen with maps or other static models (Jacobson, 1992; Neves and Camara, 1999).

10.3.2 Virtual Reality

VR is a human-computer interface in which the computer creates a three-dimensional, sensory immersing environment that interactively responds to and is controlled by the behaviour of the user (Pimmentel and Teixeira, 1995). Its characteristics are response to user actions, real-time 3D graphics and a sense of immersion. By using multimedia, users can gain real time response to their actions by graphic and sound in a virtual environment. An example is that as a 'person' walks along a street, the sound heard by that person would change continuously based on their location relative to the sound source. The richness of this experience facilitates a users' learning and understanding (Pont, 1993).

As with GIS, links between VR and the Internet have been developin
Reality Modelling Language (VRML) is a standard format for the web (Rohrer a
1997) that lets you quickly build virtual worlds incorporating 3D shapes, anin
sound effects. Since 1994, the most important developments in VR have no
technologies, but in the adoption of VR technologies and techniques to
productivity, improve team communication, and reduce costs (Brooks, 1999).

10.3.3 Internet

In the last decade of the twentieth century, the Internet emerged as a new inforn
communication technology. At the end of 1999, 1 in 5 households in the UK ha
access compared with 1 in 20 only 2 years earlier (Corrigan and Joyce, 2000).
a more efficient way for people to access information and disseminate their opi
not restricted by time or physical distance. Corrigan and Joyce (2000) and Cra
demonstrate the usefulness of the Internet to improve the productivity of publi
and contact with government representatives. Local planning authorities are be,
realise the potential of the Web as a communication device. A number of exam
where local authorities have placed important planning documentation such as
Plans and Development Plans on the web for public consultation (Carver and
1999).

10.3.4 GVIS

It is a widely held view that integrating GIS, VR and Internet technologies ca
greater and more effective participation in planning activity and thereby stren
democratise the process. Research in this field is attracting considerable attentio
Centre of Advanced Spatial Analysis (CASA) of UCL and the urban simulatic
Los Angles (Batty *et al.,* 1998; Dodge *et al.,* 1998; Jepson *et al.,* 1996). Altho
demonstration systems suggest that the technologies exist to provide functions
participation the published literature is notable for the absence of formal the
design of this type of systems. A lack of theory may undermine the developmen
and inhibit longer-term progress through rigorous evaluation. Formal theory
explain the success (or failure) of these systems and better understand
impediments to future system. An objective of the University of Salford rese
develop a robust framework for the design of GVIS.

10.4 DESIGN ISSUES 2: LEARNING SYSTEMS AND GVIS

10.4.1 Learning System Theory

Learning can be defined as the synthesis and analysis of information obtaine
communication and interaction. It can be argued that urban planning is a learnir
as it is information-rich, complex and benefits from stakeholders sharing
understanding. The exchange of information and ideas between stakeholders
informal learning environment. As such, the planning process can be consider
the framework of a theory of learning. Of the many theories of learning, Kelly's
construct theory' (Kelly, 1955) is an appropriate approach for urban planning

et al., 2001). Kelly states that people look at their world through patterns that they construct and try to fit to the real world. Without these patterns the world would make little sense to people. Patterns are constructed based on an individual's experiences, *i.e.* personal constructs. People change and revise their patterns in order to explain better their view of the world. It is noted that personal experiences include interaction with both tangible and intangible features of the world (Kelly, 1955).

To relate the learning theory to urban issues, personal constructs are formed by making sense of our direct experiences of life in the city, or indirect experiences through newspapers, books, TV and other informing media. The social interaction also lead to the building of personal constructs. Furthermore, when social interactions take place with a group, it is possible for us to envision new and more creative ways of dealing with a problematic situation by actively considering alternative constructs (Figure 10.1).

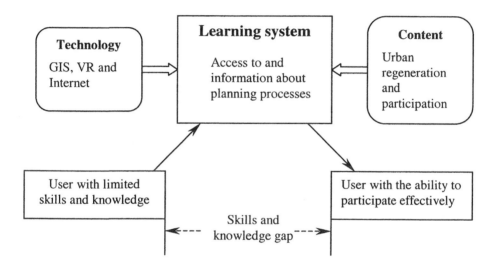

Figure 10.1 Effect of a learning system to enhance participation in planning.

Three aspects of the learning system are needed to enhance participation in the planning process, namely access, analysis, and comprehension (Zhang *et al.*, 2001). The current system to gain aaccess to information for a particular proposal can be a burden to the public because the time involved may be incompatible with their lifestyle or incur a financial penalty due to loss of earnings. Furthermore, gaining access does not improve matters unless the information is presented in a way that they can comprehend. Experimenting with alternative scenarios is another essential part of the learning system. To fully understand the planning issues, people need to analyse the information they get. The alternatives also need to be analysed and evaluated. In order to do that, public need tools to interact and refine the information. The tools may not be as complicated as the ones for the professionals but at least they can achieve some analysis functions. It has been postulated that allowing people to analyse planning proposals followed by debate between public and other stakeholders can lead to greater consensus in the final plan.

It has been observed that many people find it difficult to participate effectively in planning systems because they lack the necessary skills and knowledge (Hamilton *et al.*, 2001). A learning system could be built to bridge the skill and knowledge gap identified

(Figure 10.1). On the one side, the planning process and participation issue
content' of the system. On the other side, technologies like GIS, VR and Intern
functions that are needed to build a learning system.

10.4.2 The Strengths and Limitations of Technologies to Enhance Learning

In the light of the learning system theory, the strengths and limitations of each t
in each of the aspects of the learning system have been evaluated (Figure 1
allows people to process information and detect spatial patterns and relationshi
classified as high in terms of analysis. It is not rated as highly for
comprehension. VR is classified as high for comprehension and the Internet pi
effective way for people to access information.

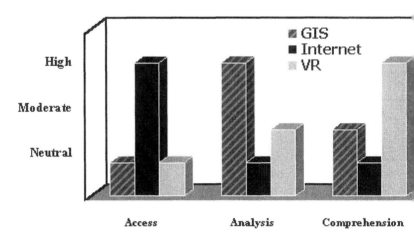

Figure 10.2 GIS, VR and Internet in Information Learning System (Source: modified from Ha∎
2001).

10.5 PROTOTYPE DESIGN AND EVALUATION

This chapter has identified design issues associated with the technologies an
systems methodology. It follows that a prototype is designed to combine the stren∎
technologies and implement features of a learning system. The research prototyp
designed around a common geospatial database for the Chapel Street Regeneratie
Salford.

Chapel Street is the main thoroughfare through the city of Salford (Figur
is also one of the main approaches to the centre of Manchester. Over the past t∎
Chapel Street has declined as a commercial and retail centre and in 1998 the Ci∎
launched a major regeneration project. The project is funded by decline ∎
private sector organisations and involves residential and business communities.

The Chapel Street case study provides a complex platform for the d∎
evaluation of a learning system-based GVIS. The prototype will be developed
with regeneration plans and is intended to complement existing activity rather th∎
it. The main benefit of using an established urban regeneration project is the∎

compare old and new approaches to participation. Before the prototype can be developed, however, it is necessary to identify those aspects of the planning process that it will address.

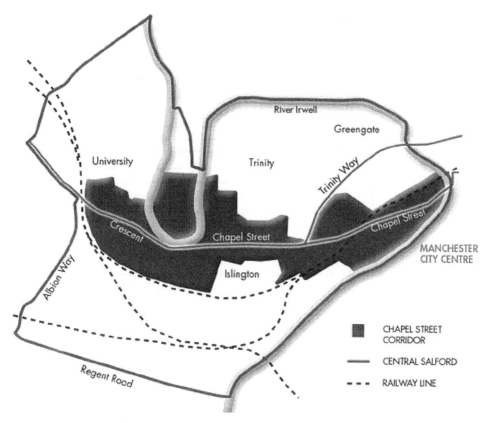

Figure 10.3 Chapel Street Regeneration Corridor, Salford, UK.

10.5.1 Urban Planning System: User Needs

In keeping with the aim of this research the planning process was studied for key stages which represent the best opportunities for the use of GVIS to support public participation. Kammeier (1999) suggested that planning support systems should support clarifying the planning options, simulating alternative proposals, assessing shortlisted projects and implementing the decision. These coincide fairly closely with Skeffington's observations (1969) that the main opportunities for public participation in a local plan are at the stages of surveys of facts, developing planning options and discussing favoured proposals. If these proposals are mapped to Yeh's model of the planning process then 4 stages can be identified which are most conducive to public participation (Figure 10.4). By comparison GIS is most useful whilst analysing the existing situation, and developing and selecting planning options because of the need for spatial analysis in these stages. The stages of modelling the existing situation and developing and selecting options offer greatest potential for using VR because it facilitates presentation and interaction.

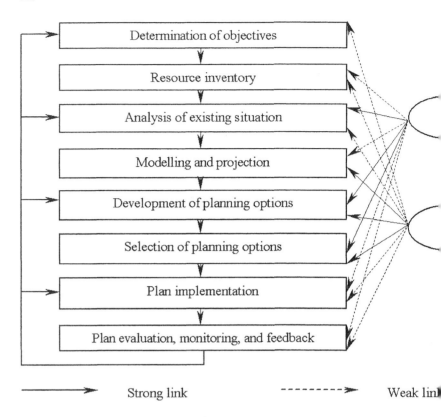

Figure 10.4 GIS, VR and public participation in planning process (modified from Yeh, 19'

Combining these assessments allows us to identify 2 stages in the plannir that are best suited to developing and testing a GVIS prototype, namely d planning options and selecting planning options (Figure 10.5).

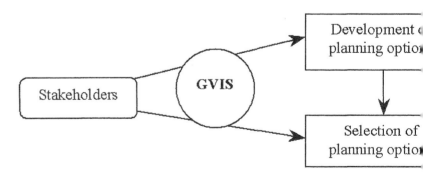

Figure 10.5 Planning stages for developing and evaluating a GVIS.

10.5.2 Learning System Components

The prototype has been designed as 3 modules that each focus on an element of the learning system.

10.5.2.1 Analysis Module

The analysis module uses spatial analysis functions to process data and support queries such as *where is...?*, *what is at location...?* and *what if...?*. In particular the module will develop decision support functions such as multi-criteria analysis.

10.5.2.2 Comprehension Module

A 3D model of Chapel Street Corridor has been built based on the same geo-spatial database of the analysis module. In this module, user can navigate and query the planning area. Main features are the provision of functions to switch between before and after views of the area based on possible scenarios and the geo-referencing of multimedia information such as video clips and panoramic photographs. The module is also designed to improve communication between stakeholders by allowing them to comment on various aspects of the regeneration plan, attach those comments to a visual object and to retrieve the comments of other participants.

10.5.2.3 Access Module

The module is mainly about the access of the planning related information through the Internet and/or Intranet. Users are allowed to access the information via a popular Internet browser. All the functions of the former two modules would be transplanted/linked in the user interface that means uses can access them by Internet or Intranet.

10.5.3 Evaluating GVIS as a Learning System

The question of whether or not new technologies improve complex tasks such as urban planning is often very difficult to answer. The evidence is usually qualitative, may be contradictory and rarely allows the system developer to make an objective evaluation. In particular it is difficult to assess the impact of combined technologies. Implementing learning system theory in the design of a GVIS prototype enables the research to reduce the system into three objectives. These objectives transcend the technologies and therefore provide a useful frame of reference for evaluating combined GVIS. Failure to achieve one of objectives can be used to focus further research and development of the prototype system.

10.6 SUMMARY

Stakeholders, including the public, are being encouraged to participate
regeneration. To do so they need to be able to contribute to and benefit from th
process. Recent studies have created virtual environments as a mechanism to fa
communication of planning information and postulate on the near-future app
cityscapes. At the same time community-based GIS researchers have argue
technologies can help to secure active participation of leading individuals and
early stages in the planning process and numerous authors promote the Int
forum for interaction between participants. In this chapter the authors have pre
case that integrating GIS, VR and Internet technologies can increase partic
planning process. More importantly, applying learning systems theory to the
GVIS creates a framework that can be used to optimise participation.

10.7 REFERENCES

Al-Kodmany, K., 1999, Using visualization techniques for enhancing public pa
in planning and design: process, implementation, and evaluation. *Lands*
Urban Planning, **45**, pp. 381-392.
Alterman, R., Harris, D. and Hill, M., 1984, The impact of public partici
planning. *Town Planning Review*, **55**(2), pp. 177-196.
Barlow, J., 1995, *Public Participation in Urban Development: the European E*
(London: PSI).
Batty, M., 1993, Using geographic information systems in urban planning a
making. In *Geographic Information Systems, Spatial Modelling and Policy E*
edited by Fischer, M. and Nijkamp, P., (New York: Springer-Verlag), pp. 51
Batty, M., Dodge, M., Doyle, S. and Smith, A., 1998, Modelling virtual enviro
Geocomputation: A Primer, edited by Longley, P., Brooks, S.M., McDonne
MacMillan , B., (Chichester: John Wiley & Sons), pp. 139-161.
Bickerstaff, K. and Walker, G., 2001, Participatory local governance and
planning. *Environment and Planning A*, **33**, 431-451.
Brooks, F.P., 1999, What's real about virtual reality?. *IEEE Computer Gra*
Applications, **19**(6), 16-27.
Carver, S. and Peckham, R., 1999, Internet-based applications of GIS in pla
Geographical Information Systems in Planning: European Perspectives,
Stillwell, J., Geertman, S., and Openshaw, S., (New York: Springer-Verlag)
390.
Corrigan, P. and Joyce, P., 2000, Reconnecting to the public. *Urban Studies*, 3
1771-1779.
Cowan, R., 1998, The people and the process. In *Introducing Urban*
Interventions and Responses, edited by Greed, C., Roberts, M., Longman.
Craig, W. J., 1998, The Internet aids community participation in the planning
Computers, Environment and Urban Systems, **22**(4), pp. 393-404.
DETR, 1998, *Guidance on Enhancing Public Participation in Local Go*
(London: The Stationery Office).
Dodge M., Doyle S., Smith A., and Fleetwood S., 1998, Towards the virtual c
Internet GIS for urban planning, paper presented at the *Virtual Re*
Geographical Information Systems Workshop, Birkbeck College, London.

Dollner, J. and Hinrichs, K., 1998, An object-oriented approach for integrating 3D visualization systems and GIS. *Computers and Geosciences*, **26**(1), pp. 67-76.

Douven, W., Grothe, M., Nijkamp, P., and Scholten, H., 1993, Urban and regional planning models and GIS. In *Diffusion and Use of Geographic Information Technologies*, edited by Masser, I. and Onsrud, H., NATO ASI Series, (Dordrecht: Kluwer Academic).

Egenhofer, M. and Kuhn, W., 1999, Interacting with GIS. In *Geographical Information Systems*, 2nd., edited by Longly, P., Goodchild, M. F., Maguire, D. J., Rhind, D. W. , (New York: John Wiley & Sons), pp. 401-412.

Frank, A., and Mark, D., 1991, Language issues for geographical information systems. In *Geographical Information Systems: Principles and Applications,* 1st., edited by Maguire, D. J., Goodchild, M. F., Rhind, D. W., (London: Longman), pp. 227-237.

Geddes, P., 1905, *Civics: as applied sociology*. Sociological papers 1904, (London: Macmillan).

GeoWorld, 1999, Industry trends: GIS industry outlook 2000 - the birth of a new millennium. *GeoWorld,* December. http://www.geoplace.com/gw/1999/1299/1299ind.asp.

Goodchild, M.F., 1992, Geographical information science. *International Journal of Geographical Information Systems*, **6**(1), pp. 31-45.

Hamilton, A., Trodd, N., Zhang, X., Fernando, T., Watson, K., 2001, Learning through visual systems to enhance the urban planning process. *Environment and Planning B: Planning and Design,* Oct., 2001.

Ingram, R., 1998, Building virtual worlds: a city planning perspective. *Online Planning Journal,* http://www.casa.ucl.ac.uk/planning/articles3/vcity.htm.

Jacobson, B., 1992, The ultimate user interface. *Byte,* **17**, pp.175-182.

Jepson, W., Liggett, R. and Friedman, S., 1996, Virtual modelling of urban environments. *Presence,* **5**(1), pp. 72-86.

Kammeier, H., 1999, New tools for spatial analysis and planning as components of an incremental planning-support system. *Environment and Planning B: Planning and Design,* **26**(3), pp. 365-380.

Kelly, G. A., 1955, *The Psychology of Personal Constructs,* 2 Vols., (New York: Norton).

Klosterman, R., 1999, The what if? Collaborative planning support system. *Environment and Planning B: Planning and Design,* **26**(3), pp. 393-408.

Maguire, D. J., 1991, An overview and definition of GIS. In *Geographical Information Systems: Principles and Applications,* 1st., edited by Maguire, D. J., Goodchild, M. F., Rhind, D. W., (London: Longman), pp. 9-20.

McConnell, S., 1981, *Theories for Planning*, (London: Heinemann).

Medyckyj-Scott, D. and Hearnshaw, H., (editors), 1993. *Human Factors in Geographical Information Systems*, (London: Belhaven Press).

Myers, J., Martin, M. and Ghose, R., 1995, GIS and neighbourhood planning: a model for revitalizing communities. *Journal of Urban and Regional Information System Associations*, **7**(2), pp. 63-67.

Nedovic-Budic, Z., 1994, Effectiveness of geographic information systems in local planning. *Journal of the American Planning Association*, **60**(2), pp. 244-263.

Nedovic-Budic, Z., 1998, The impact of GIS technology. *Environment and Planning B: Planning and Design*, **25**(5), pp. 681-692.

Nedovic-Budic, Z., 2000, Geographic information science implications for urban and regional planning. *URISA Journal*, **12**(2), pp. 81-93.

Neves, J. N. and Camara, A., 1999, Virtual environments and GIS. In *Geo Information Systems,* 2nd., edited by Longley, P.A., Goodchild, M.F., Mag and Rhind, D.W., (New York: John Wiley & Sons), pp. 557-65.

Peng, Z., 1997, An assessment of the development of Internet-GIS, presented at *User Conference.*

Pimentel, K. and Teixeira, K., 1995, *Virtual Reality: Through the New Look* 2nd., (New York: MacGraw-Hill).

Pont, P., 1993, Applied virtual reality. In *Virtual Reality in Engineering,* Warwick, K., Gray, J. and Roberts, D., (London: Institute of Electrical Engi

Roberts, P. and Lloyd, G., 1999, Institutional aspects of regional planning, ma and development: models and lessons form the English experience. *Enviror Planning B: Planning and Design,* **26**(4), pp. 517-531.

Rohrer, R. and Swing, E., 1997, Web-based information visualization. *IEEE Graphic and Application,* pp. 52-59.

Rydin, Y., 1999, Public participation in planning. In *British Planning –50 Years and Regional Policy,* edited by Cullingworth, B., (London: Athlone), pp. 18

Sarjakoski, T., 1998, Networked GIS for public participation emphasis on utiliz data. *Computers, Environment and Urban Systems,* **22**(4), pp. 381-392.

Skeffington Report, 1969, *People and Planning.* (Report of the committee participation in planning), (London: HMSO).

Weiner, D., Warner, T., Harris, T., Levin, R., 1995, Apartheid representations i landscape: GIS, remote sensing and local knowledge in Kiepersol, Sou *Cartography and Geographic Information Systems,* **22**(1), pp. 30-44.

Whittick, A., 1974, *Encyclopaedia of Urban Planning.* (New York: McGraw-H

Yeh, A. 1999, GIS and Urban Planning. In *Geographical Information Syste* edited by Longley, P., Goodchild, M. F., Maguire, D. J., Rhind, D. W., (N John Wiley & Sons), pp. 877-888.

Zhang, X., Trodd, N. and Hamilton, A., 2001, Geographic & visual informati (GVIS) to support advanced urban planning: a research framework. In *Proc the First International Conference of Postgraduate in the Build an Environment,* pp. 622-633.

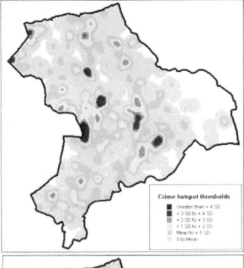

Plate 1 The map applies incremental standard deviations above the mean to define crime hotspot legend thresholds. The crime data mapped is robbery (June-August 1999). The map was produced using a bandwidth of 207m (K order 5).

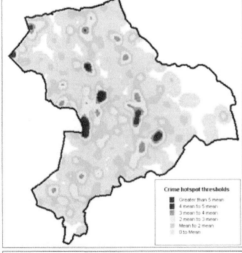

Plate 2 An incremental mean approach for defining legend thresholds provides an appealing method for describing concentrations of crime. The crime data mapped is robbery (June-August 1999). The map was produced using a bandwidth of 207m (K5).

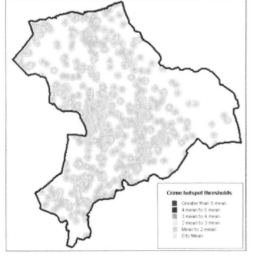

Plate 3 Gi* LISA statistic output (grey areas) derived from the robbery (117m (K2) bandwidth) quartic kernel density estimation surface. The LISA output matches very closely to the ranges above the 3 mean threshold, and is useful in adding definition to our original continuous surface crime hotspot map, indicating the level at which hotspots can be more clearly distinguished against other levels of crime concentration.

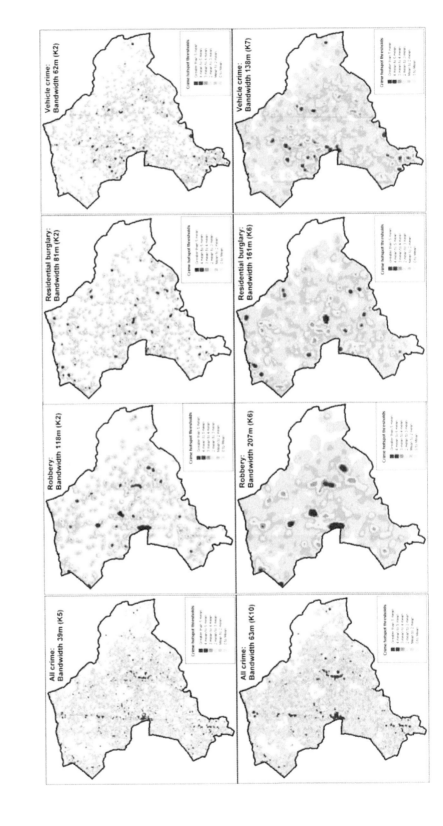

Plate 5 Canary Wharf modelled in Canoma.

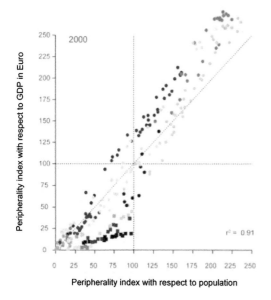

Plate 6 Canoma model of Rosemary Gardens, displayed in Viewpoint.

Plate 7 Correlation of *Peripherality with respect to population* against *Peripherality with respect to GDP* (NuTS 2).

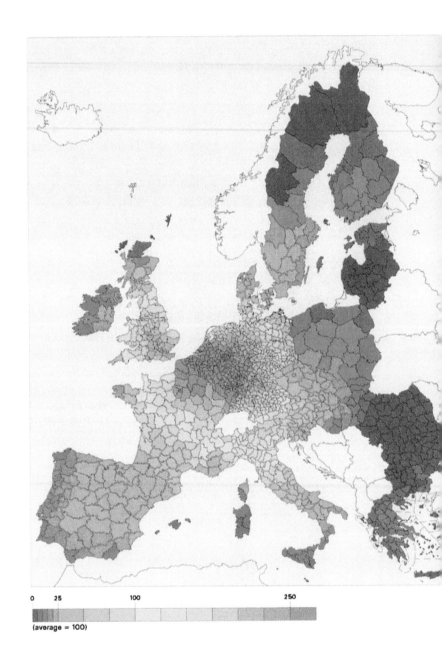

Plate 8 Peripherality index with respect to GDP by lorry (NUTS 3).

11

Using measures of spatial autocorrelation to describe socio-economic and racial residential patterns in US urban areas

Andrea I. Frank

11.1 INTRODUCTION

Residential patterns are investigated using an exploratory approach with GIS-based thematic maps and spatial statistics. The research aims to contribute to the debate whether geopolitical fragmentation enhances spatial segregation of households based on racial and socio-economic characteristics. Using 1990 US census data for three key variables (income per capita, percent black and percent hispanics), measures of spatial association serve as quantitative index to describe the level of homogeneity or heterogeneity of residential neighborhoods. Statistical results are mapped and the correlation of significant homogeneous clusters with political entities is assessed. The results show considerable variation of household cluster sizes depending on the geopolitical structure as well as the population composition of the urban area. Also, cluster sizes and locations are correlated with municipal boundaries in some of the areas; this supports the proposition that the location of segregated communities is influenced by political and policy frameworks.

11.2 BACKGROUND

Within the fabric of the city, households of similar socio-economic, racial or ethnic background tend to cluster spatially forming distinct neighbourhoods and communities. Human geography and urban planning have a long tradition in observing and analysing the spatial pattern of residential arrangements and the change of such arrangements in time and space as a means to develop and test explanatory models of social interaction and urbanisation processes. It is from such observations that theories of human ecology and the concentric *zonal* model (Burgess, 1925) of urban structure were developed in the 1920's by scholars from the Chicago School (*e.g.* Park *et al.*,1925; Park, 1936; Hawley, 1950). In the context of movements such as sustainability and social justice, knowledge of

patterns of residential differentiation is vital to raise awareness of the spatial is
disadvantaged households lacking access to health care, jobs and transportation.

Recent research (*e.g.* Foster, 1997; Pendall, 2000) suggests that the spati
of social and racial groups in the US is not necessarily based on free choi
induced and facilitated in part by local government policies on land use, zoning
and restrictive policies like minimum floor space requirements for new homes. *f*
of the influence of local governments on the residential mosaic we might expe
that the socio-economic structure and patterns mirror the underlying spatial
structure. In other words, household characteristics might be changing significa:
crossing jurisdictional boundaries. This study continues the tradition of researcl
residential patterns looking specifically at linkage of segregation and ge
structures. The objective is to a) evaluate and compare different patterns and
residential segregation in US urban areas and b) test the hypothesis of the asso
political structures with patterns of residential segregation.

The paper consists of three parts. The first part briefly reviews theories
residential segregation and the potential influences of geopolitical structures
patterns. The second part discusses the analysis approach, which is based c
statistics and pattern analysis techniques. Spatial statistics is a helpful tool to a
distribution of household types quantitatively using spatial association meası
level of integration (co-location) and segregation of socio-economic and racial
analysed using multivariate spatial correlation, whereas the correspondence of ju
boundaries with the extent of statistically significant patches of similar hous
evaluated through visual assessment.In the third part, the analysis is applied to
urban areas with different political structures. Spatial clusters of socio-
indicators are evaluated and compared with the areas of local political entities.

11.3 URBAN RESIDENTIAL MOSAIC AND POLITICAL STRUCTURE

Residential segregation or congregation (depending one's point of view) is
based on socio-economic status, ethnicity, race, and to a lesser degree life-cyc
lifestyle, and age (Knox, 1994). A range of theories on residential segregation e
one hand, the phenomenon is seen as a result of the complex dynamics
interaction. Lasting social contacts are based on commonalties, such as hobbies
and cultural values. Commonalties often derive from similar educational bacl
occupations and socio-economic status. The shared experience of university life
the same issues at work often forms the basis of a friendship. On the other hai
interaction is greatly facilitated by physical closeness (Johnston, 1982); this
significant proportions of the population despite technology that renders
increasingly obsolete such as telecommunications, improved personal mobilit
Internet (Dodge and Kitchin, 2000). Thus, it seems reasonable that people
proximity of others with whom they share common interests and values. Be
residential segregation go beyond the convenience of proximity. Suttles (1
example, argued that residential segregation leads to minimisation of inter-grouף
maximisation of political voice, and an increased degree of social contrc
segregation helps to preserve cultural values, provides a haven of moral, spii
practical support and reduces the impact of discrimination (*e.g.* Boal, 1976).

Residential segregation is a matter of fact in US metropolitan areas; more iı
is therefore *where* certain groups locate within the metropolitan fabric and *how*
neighbourhoods relate to each other. In general, the location of different social a
groups within a metropolitan area is determined through an approximate matc

value and resources available to a group. This has led to the development of a generic model of urban structure with poor, black or minority urban centres and wealthy (white) urban fringes. Yet, the urban residential mosaic is not static and local conditions vary. Downtown neighbourhoods are being gentrified by young urban professionals displacing less affluent former residents, for example, while inner city industrial land is converted to housing and so forth.

Residential self-sorting may be facilitated by municipal and school district boundaries (Weiher, 1991). The statement resonates the view offered by public choice theory (*e.g.* Tiebout, 1956). According to Tiebout (1956), households will choose and locate in the municipality, which offers the closest match to the households preferred level of services, costs and environmental conditions. Thus, municipalities will over time attain a population that is relative homogeneous by taste. A factor analysis of demographic variables led Heikkila (1996) to conclude that the various municipalities of Los Angeles indeed resemble Tieboutian clubs of relative homogeneous population groups. As households with similar characteristics often share taste and preference, we may assume relative high levels of homogeneity by household type, economic class, and race as well.

Municipalities are not passive players in this game but compete like 'municipal firms' for residents. To the extent that municipalities influence land values and access to their territory by means of policy and zoning, they have a direct impact on the socio-spatial pattern of the metropolitan area. It may be fiscally advantageous for municipalities to specialise. Newly incorporated cities and municipalities, that cater to specific population groups (*e.g.* the retirement communities of the Sun Belt), or land uses (*e.g.* City of Industry, CA) are examples for this specialisation. In effect municipalities and local governments act as territorial containers. Through specific local government legislation certain land uses or population groups can be offered preferential treatment or denied access. The size of jurisdictions may be of importance for the facilitation of the segregation process. Access control becomes more difficult, the larger the area. Hence, larger jurisdictions are less likely to be highly homogeneous by class, race or along any other dimension (Madison cited in Sack, 1986: 147). From this we might conclude that in urban areas with lots of small jurisdictions, segregation might be more easily attained than in urban areas with large jurisdictions.

To summarise, the reasons for residential segregation are diverse ranging from different tastes and preferences, racial prejudice and discrimination to economic barriers. There is some indication that the self-sorting process of households is facilitated by the differential conditions in local governments. As there is only a certain proportion of an urban population that fits into each grouping (*e.g.* the Asian community, the wealthy, the black entrepreneurs, *etc.*), spatial containers and jurisdictions ideally should be small for better fit. It is expected that in urban areas with many local government entities: (i) segregation patterns reflect the small-scale political structure; and (ii) boundaries of clusters of similar households are often matching local government boundaries. In contrast, in urban areas with few government entities (fewer choices), clusters of households with similar race, ethnicity or social status are likely to form larger clusters that not necessarily match up with the boundaries of jurisdictions.

11.4 SPATIAL PATTERN ANALYSIS

Describing and comparing the residential mosaic of different cities is not a simple task. In a research context, social area analysis (a technique based on factor analysis) has successfully been used to generate generic descriptions of the socio-spatial structure of a wide range of cities. The results showed that residential differentiation is primarily dependent on three dimensions: socio-economic status, life cycle and ethnici█ households. The technique is difficult to interpret, however, and critics voicec over the dependency of analysis results on the underlying spatial structure. █ phenomena in space tend to be interdependent (*e.g.* Tobler, 1970; Gou█ Households with similar characteristics (socio-economic status, age, ethnicity likely to be located near each other, and are thus in a statistical sense dep autocorrelated. This spatial association of household locations casts further do validity of social area analysis results as variable autocorrelation may bias the the underlying factor analysis.

Scholars have been aware of the tendency of spatial autocorrelation in g data (Tobler, 1970; Gould, 1970) and have developed a variety of approach with the phenomenon. One response is to determine the degree of interdepen measures of autocorrelation and allow for the necessary corrections to avoid errors (Cliff and Ord, 1981; Anselin, 1996). A second approach is autocorrelation as a general property found in variables of any system, ma█ natural, observed over time (temporal autocorrelation) and/or space autocorrelation) (Legendre, 1993). The spatial associations of variables a█ valuable indicators of life supporting processes and organisation and spatial a measures can be viewed as a quantitative index for the description of spatial pa processes (Cliff and Ord, 1981; Goodchild, 1986; Legendre, 1993; Shen, 1994).

Patterns in space can be defined based on the heterogeneity or homo█ adjacent observations and the size and arrangement of patches of similar relation to each other. Spatial association measures, which determine the heterogeneity or homogeneity of variables in space, can therefore be used to the degree of segregation of households in an urban area.

Overall two aspects of urban residential patterns are investigated. On *characteristics* of today's socio-economic pattern in urban areas are explore other hand, the *correlation* between boundaries of local government entitie spatial arrangements and size of segregated residential clusters is evaluated. In step analysis, socio-economic patterns are first visualised on thematic maps a█ described both qualitatively as well as quantitatively via the level of spatial assc three key indicators of residential differentiation: income, race and ethnicity. Th of spatial autocorrelation indicates the level of segregation for each variable spatial association indicates that similar values of a variable under study are c█ space; this would be an indicator of segregation. Negative spatial autoc indicates that similar values of a variable under study are dispersed to a deg unlikely under randomisation. No spatial autocorrelation means the pattern o█ likely to be random. Third, a visual inspection of map overlays of significant █ segregated households with municipal boundaries addresses the questio correlation of the political structure and identified residential segregated clu█ analysis compares residential patterns of eight different urban areas.

The analysis is conducted using two software applications working in tan loose coupling approach (Anselin, 2000): SpaceStat™ (Version 1.90), a spatia software and ArcView™ (Version 3.1), a vector-based GIS from the Envi Systems Research Institute (ESRI). Geographic co-ordinate data of enumerati█ centroids and topology are exported for use in a spatial weights matrix and analysis results are fed back into the GIS for visualisation.

---

[1] In the US context White, Black or Asian, represent racial population categories, whereas Hispa█ ethnic heritage; Hispanics may be of any race.

11.4.1 Analysis Unit

The census designated Urbanized Area (UA) rather than the Metropolitan Statistical Area (MSA) is used as primary analysis unit. MSAs, which are based on county boundaries, frequently include large areas of undeveloped land. This could distort the residential pattern analysis and was thus found unsuitable for the analysis. The census designated UA was introduced to help distinguish between urban and rural land use and is a standard enumeration unit since 1960. A UA comprises of a minimum of 50,000 persons, one or more central places and surrounding urban fringe territory. Fringe territory must have a population density of at least 1000 persons/square mile to be included (GARM, 1994). As a consequence, the UA describes best the extent of a densely populated area with a territorial definition that is largely independent of administrative boundaries such as county or municipal boundaries.

Case study UAs were selected for comparison along two aspects: political structure and population mixture. The political structures of the selected UAs' range from very simple with a few governments to ones that stretch over several states and consist of 30+ local governmental entities. The selected UAs are located in different regions of the US displaying different racial and ethnic composition. Southwestern UAs (*e.g.* Albuquerque) have a high level of Hispanics in the overall population, whereas Eastern UAs (*e.g.* Richmond) house a higher percentage of black population (Table 11.1). The analysis was conducted on midsize urban areas with a population between 250,000 and 1,000,000.

Table 11.1 Urbanized Area complexity/fragmentation and racial/ethnic composition

UA Name	# of Gov't entities / fragmentation	Racial and Ethnic Composition					
		% White	% Black	% Native	% Asian	% Other	%Hisp Origin
Tucson	4 / low	78.6	3.4	1.8	2	14.2	*19.3*
Albuquerque	6 / low	77.9	2.8	2.75	1.55	15.0	*36.25*
Richmond	4 / low	68.5	29.4	0.2	1.6	0.3	*1*
Oxnard-Ventura	9 / med-low	78.2	2.8	0.7	5.7	12.6	*26.3*
Oklahoma City	28 / med-high	80.1	11.7	4.2	2.1	1.9	*3.7*
Harrisburg	38 / high	84.6	12.5	0.2	1.5	1.2	*2*
Akron	40 / high	86.9	11.7	0.2	1	0.2	*0.6*
Providence-Pawtucket	38 / high	90.75	4.05	0.4	2	2.8	*5*

11.4.2 Data

For the analysis, 1990 US census data was extracted from ArcData Online[2]. The database provides geographic and attribute data. At the time of download data were based on the Tiger® 1995 products of the US census. Block groups serve as the spatial units for the representation of the residential pattern. Block groups are the smallest enumeration unit for which the US Census provides socio-economic and demographic data on a general basis. Boundaries of independent local government units (towns, townships, parishes, villages, cities), counties and states are used to describe the geopolitical structure of the areas. Three variables were used to portray the socio-economic pattern:

---

[2] ArcData Online from ESRI (Environmental Systems Research Institute) can be accessed via the Internet from URL: http:www.esri.com/data/online/tiger.

- Income per Capita (INPRCAP),
- Percent Blacks (PERBLACK),
- Percent Hispanics (PERHISP).

11.4.3 Spatial Statistics

The evaluation of the level of spatial association of variables values from enumeration districts is influenced by the definition of nearness or proximity. F oriented data models like the TIGER® census files, data are aggregated enumeration district and conceptualised at discrete locations. Thus, in order to spatial autocorrelation of income per capita, for example, the aggregate val census block group is compared to the equivalent income per capita neighbouring block groups.

For this study, neighbourhood is defined as a) *dichotomous*, assum boundaries between ethnic, racial or socio-economic enclaves are relative sharp and b) *distance-dependent*, assuming that segregated communities are not infini In other words, every spatial unit whose centroid is within a specified distance spatial unit's centroid is considered a neighbour and every spatial unit with outside the specified distance is not. The concept of this neighbourhood relat depicted in Figure 11.1. For block group **A** (centre) and the distance **r**, only group is defined as neighbour (dark grey polygon); for distance **2r**, 21 block g into the neighbourhood range.

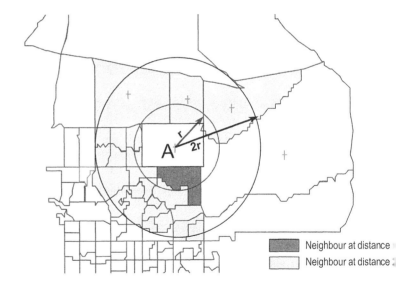

Figure 11.1 Distance-based Neighbourhood Relationship.

The neighbourhood definition results in small block groups having more n and large rural block groups at the fringe of the analysis regions being elimina is desirable. Perry (1929) has defined the spatial extent of a neighbourhood unit enclosed by arterial, high-volume roads and served by a school. Considering th low density of US urban areas and the block group size, an idealised neighbo assumed to be not larger than a circle with r=2000m (approx. 4.8 square m distance r=2000m is also greater than the average distance between centroids for all UAs, which is important methodologically as distance cut-offs smaller than the average will lead to a non-continuous analysis surface and missing values in the spatial weights matrix.

The level of spatial autocorrelation is expressed by the global Moran's *I* coefficient, which is calculated in SpaceStat[TM] as outlined in Equation 11.1, whereby *N* depicts the number of observations, w_{ij} represents an element in the spatial weights matrix corresponding to an observation pair *i,j*, while x_i, x_j are observations for locations *i* and *j* (with mean μ) and S_0 is a scaling constant of the sum of all observation pairs (Eq. 11.2).

$$I = \frac{N}{S_0} \frac{\sum_i \sum_j w_{ij}(x_i - \mu)(x_j - \mu)}{\sum_i (x_i - \mu)^2} , \quad \text{with} \tag{11.1}$$

$$S_0 = \sum_i \sum_j w_{ij} \tag{11.2}$$

For any number of polygons *N,* a Moran's *I* coefficient approaching +1 indicates strong positive autocorrelation and -1 indicates negative autocorrelation. A coefficient of $-1/(N-1)$ indicates zero autocorrelation or a random pattern. The Moran's *I* is a global measure of autocorrelation, which is strictly speaking only valid under the assumption of spatial stationarity. As the assumption of spatial stationarity is unlikely to hold for regions with several dozen and more spatial observations, a local measure of spatial association - the local Moran's I_i (Equation 11.3) - was introduced to detect local instability and nonstationarity. Values z_i, z_j represent deviations from the mean, while the summation over *j* includes only neighbouring values $j \in J_i$. The average of all I_i will equal the global Moran's *I*. Extreme local Moran values represent outliers with strong deviation from the mean of the overall sample.

$$I_i = (z_i / m_2) \sum_j w_{ij} z_j, \quad \text{with } m_2 = \sum_i z_i^2 \tag{11.3}$$

Moran scatterplot maps are used to depict the analysis results; they are deemed instrumental in the exploration of spatial patterns (Anselin, 1996; Anselin and Bao, 1997). A scatterplot map is based on a statistic that can be visualised as the slope of a straight line that indicates the level of global association. When the statistic is decomposed into four types of association, the lower left and the upper right quadrant indicate positive spatial autocorrelation, *i.e.* the presence of similar (low or high) values in neighbouring locations. The upper left and lower right quadrants indicate negative spatial association or the presence of dissimilar values in neighbourly locations. The results are mapped by assigning different colour schemes to values from the four quadrants. Outliers, *i.e.*, values significantly above or below the mean can be identified and shown in additional maps. Scatterplot maps together with maps of significant local indicators of spatial association are key components of the analysis.

11.5 CASE STUDY RESULTS

The residential pattern of eight different UAs were analysed and compared. As expected, significant clustering of households was discernible from mapping the racial and ethnic variables. Spatial association for the income variable is not as significant. Furthermore, patterns vary depending on the political structure and population composition. These differences are illustrated below contrasting two UAs: Tucson, with a simple structure and a large Hispanic population and Harrisburg, with a complex structure and a high percentage of blacks in the population.

11.5.1 Tucson

Figure 11.2 shows the UA boundary with a thick black line. Jurisdiction over the shared by only 4 different governments: the City of Tucson, South Tucson and within the City of Tucson, Oro Valley town in the Northwest and Pima County.

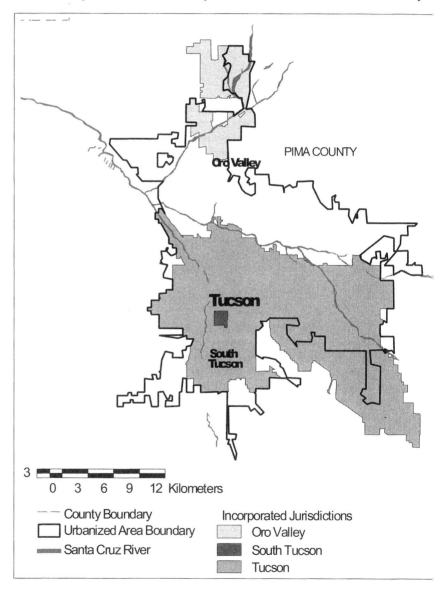

Figure 11.2 Spatial political structure of Tucson UA.

Figure 11.3 shows the location of block groups with higher or lower than average levels of Hispanic households in the Tucson UA. The scatterplot map shows that the UA is divided in two parts: one which is dominantly Hispanic (high-next-to-high values of PERHISP), and one which is dominantly non-Hispanic (low-next-to-low values of PERHISP). The large grouping of dark grey block groups in the south of the UA depicts the part that is dominated by Hispanic households. While circa 20% of the residents of the entire UA claim Hispanic origin, the proportion of Hispanics in this contiguous patch of block groups is 40 to 95% indicating significant homogeneity for ethnic origin. This is clearly not random. Block groups that are highly Hispanic next to block groups with little or no Hispanic population are shown in middle grey shades, whereas the light grey block groups depict areas for which the proportion of Hispanics is consistently below 19%.

Figure 11.4 shows again two large clusters. The cluster of wealthy households (high next to high INCPRCAP values) is located in the north of the UA – outside the city in Pima County. The per capita income in the Tucson UA in 1990 was US$ 13,105. The average per capita income for the block groups in the cluster located in Pima county is over US$ 28,000. The cluster of poorer households (low next to low INCPRCAP) is located to the south in the city of Tucson and South Tucson. The boundary between high and low-income areas is not as clear as for the ethnic variable.

Only 3.4 % of the UA population in Tucson are black; most of them live north of the Hispanic neighbourhood but within the city limits. It seems that the black community is part of mixed income and racial area that forms a buffer between the rich white and poorer Hispanic population.

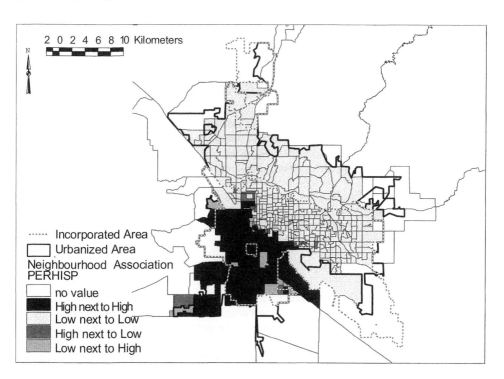

Figure 11.3 Moran scatterplot for Hispanics in Tucson (r=2000m).

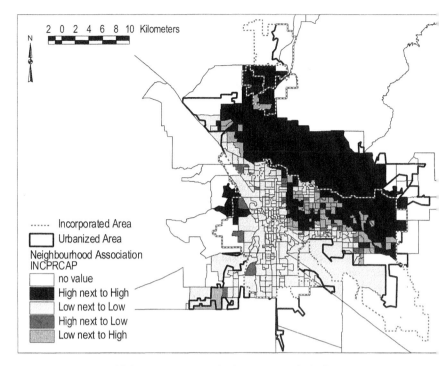

Figure 11.4 Moran scatterplot for Income per capita in Tucson UA (r=2000m).

The clustering, *i.e.* spatial autocorrelation, of households of similar eth income is confirmed by the high global Moran's *I* coefficients for both the (=0.859) and INCPRCAP (0.551) variables (see Table 11.2 below). Negative m spatial correlation of the PERHISP and INCPRCAP variables points als separation of high-income from Hispanic households. The Hispanic househo southern part of Tucson are not only Hispanic but also much poorer on average.

Maps and statistics show a stark pattern of economic and ethnic segreg clarity of the political structure reflects the large-scale political structure of Residents do not have many choices except for Oro Valley town and South Tucs is only the choice to live inside or outside the city limits of Tucson. The wealthier households to locate outside the city proper is common in US me areas and consistent with theories of white flight from redistributive urban taxes.

11.5.2 Harrisburg

The political structure of the Harrisburg UA is complex, consisting of government entities including townships, counties, towns and villages (Fig Towns and villages denoted by letters *a - r* on the map are fairly compact ar proportion of the populations lives in townships.

Overall clusters of adjacent block groups with significant homogeneous p groups (race, economic status or ethnicity) are much smaller in comparison. Fi depicts four clusters for income per capita (INCPRCAP). There are two larger c areas to the east of the central city of Harrisburg and two to the west. Clusters clearly delineated as in the Tucson UA. As a result the pattern for income is spatially autocorrelated than in Tucson, which is reflected in the lower Moran's *I* coefficient (0.206) for INCPRCAP. The pattern for PERBLACK, however, is spatially correlated as indicated by a Moran's *I* coefficient of 0.668 (Figure 11.7 and Table 11.2). Blacks are mostly confined within the boundaries of the central city of Harrisburg. This is also the poorer area. The pattern is almost a perfect reversal to that of the income pattern in Figure 11.6. Multivariate spatial correlation of the INCPRCAP and PERBLACK variables is negative with –0.50 pointing also to the spatial segregation of Blacks and high income households. The correlation of boundaries of racially segregated areas with the boundaries of jurisdictions is considerable.

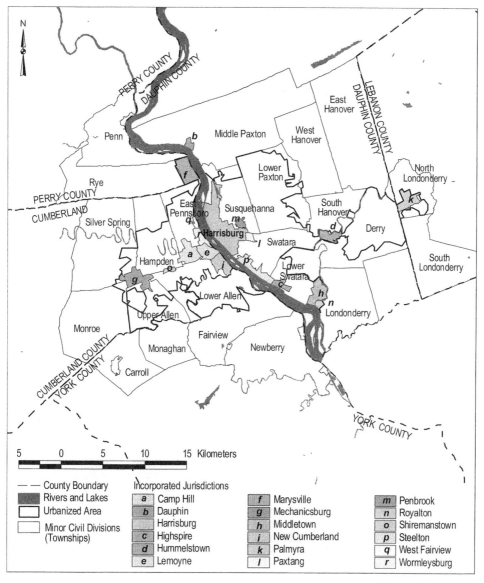

Figure 11.5 Spatial political structure of Harrisburg UA.

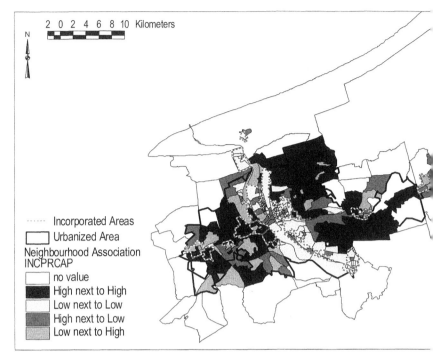

Figure 11.6 Moran scatterplot for INCPRCAP in Harrisburg (r=2000m).

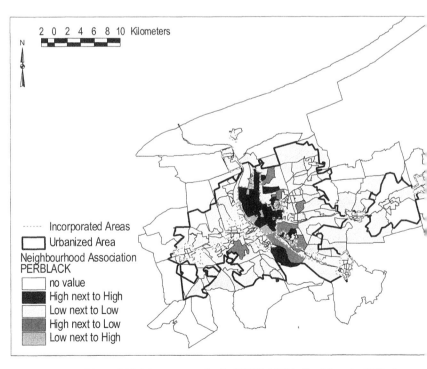

Figure 11.7 Moran scatterplot for PERBLACK in Harrisburg (r=2000m).

11.5.3 Other Results

When comparing spatial autocorrelation of the variables across the eight different case studies autocorrelation levels for the ethnic/racial variables show a significant geographical bias (Table 11.2). All five Eastern UAs show significant and in the case of Oklahoma City and Richmond very high positive autocorrelation for the PERBLACK variable. In five of the eight urban areas the autocorrelation of Hispanics is a dominant force in shaping the residential pattern. Three of these urban areas are located in southwestern states, which have a high proportion of Hispanics in the urban population. The other two UAs with a significant Moran's *I* coefficient for Hispanics are Oklahoma City (0.658) and Providence-Pawtucket (0.635). These two UAs display significant autocorrelation for PERBLACK and PERHISP.

Interestingly, the level of global spatial association for INCPRCAP is less significant and only for four of the eight test cases is the Moran's *I* coefficient greater than 0.4. One of them is in the Tucson UA, where a large cluster of high-income households is located in the north of the UA (Figure 11.4).

In two of the UAs, Richmond and Akron, the PERHISP variable approaches a zero level of autocorrelation or random distribution of households. This result may be an artefact of the overall low level of Hispanics in the population (Table 11.1). The differences in spatial structure of ethnic and racial groups is likely to be related to the overall population mixture, *i.e.* the proportion of blacks or Hispanics as part of the overall urban population.

Table 11.2 Level of Spatial Association for sample Uas.

Urbanized Area	Geogr. region	Level of spatial association (Global Moran's *I* for distance r=2000 m)		
		INCPRCAP	PERHISP	PERBLACK
Albuquerque, NM	SW	0.332	**0.772**	0.358
Oxnard-Ventura, CA	SW	0.235	**0.664**	0.334
*Tucson, AZ***	*SW*	*0.551*	*0.859*	*0.311*
Oklahoma City, OK	SE	**0.453**	**0.658**	**0.835**
Richmond, VA	SE	**0.408**	*0.190*	**0.701**
Akron, OH	NE	**0.535**	*0.050*	**0.636**
*Harrisburg, PA***	*NE*	*0.206*	*0.382*	*0.668*
Providence-Pawtucket RI/MA	NE	0.311	**0.635**	**0.547**

The resolution of the pattern, *i.e.* the size of relative homogenous groupings of households seems to vary as well. In older, northeastern urban areas groupings are smaller and more numerous, whereas in the newer southwestern areas the socio-spatial structure tends to consist of few large districts. This may be a function of the different political structures, which may influence residential choice and foster population segregation. Political structures on the whole tend to be more small-scale and fragmented in the North and Northeast of the US and large-scale in the Southwest. In any case, distinctly high-income block groups are repeatedly found outside but abutting central city limits.

Based on the small sample, no clear relationship can be determined between the level of spatial association (as in the global Moran's coefficient) and the complexity of

the political structure. It was expected that groupings of block groups (
households would be smaller in the UAs with a more complex urban political
and therefore that the Moran's coefficients would be less significant as well. In
of INCPRCAP, Harrisburg, Oxnard-Ventura and Providence Pawtucket (all with
small scale political structures) show little spatial association as expe
Albuquerque has a similarly low Moran's *I* despite a much simpler political
Reversibly, Oklahoma City has an equally high positive spatial autocorre
INCPRCAP than Tucson, despite a much more complex political struct
relationship for PERHISP and PERBLACK are also inconsistent, howe
separated by geographic region a negative relationship between the number o
entities and the level of spatial association can be shown for the Northeastern
PERBLACK and the Southwestern UAs and PERHISP. This is in suppo
reasoning that more complex structures facilitate sorting into smaller entities t
not be adjacent and therefore result in a low global spatial association.

11.6 PLANNING IMPLICATIONS AND FUTURE RESEARCH

The study revisits residential pattern research using widely available census d
spatial-statistical analysis approach. The goal was twofold: a) to investigate cont
patterns of residential segregation in US urban areas, and b) examine whether th
and size of segregated communities coincide with local political entities. This
and its results are of interest to urban planners and planning agencies for it ma
relevant insight into the location of minority households and patterns of segregat

In terms of segregation and residential patterns, the research findings ind
global spatial association for race and ethnicity tend to be higher than spatial as
of income. The degree of spatial association is somewhat dependent on the prop
racial or ethnic minorities in the area and geographically biased. In addition, m
analysis shows that locational overlap of high-income with PERBLACK or I
tends to be negative, *i.e.* Hispanics and Blacks tend to reside in more depriv
Segregation by economic status, it seems, is not as significant a factor than
ethnicity or as Harvey stated segregation by ethnicity and race tends to wea
higher income.

In many instances, cluster dimensions and location indicating similar eth
or high-income households correlate significantly with governmental enti
example in the Tucson UA, where a large group of high-income block groups
outside the city limits or Harrisburg, where the central city holds most of
population. This correlation supports the hypothesis that local government p
some influence on residential choice and land use. Further research at the loca
needed to pursue the reasons for these socio-economic boundaries and divisi
research could be in form of additional case studies to explore geographically de
trends further, or it could be in form of detailed local, ethnographic studies. As c
high-income households that coincide with jurisdictional boundaries may be ind
discriminatory development practices, an examination of the local po
discriminatory practices and regulations may be helpful.

There is value not only in the content or results, however. In contrast to
thematic mapping, the approach takes into account spatial interdepender
evaluates the level of chance at which a certain pattern may occur.
unfortunately, the issue of a predetermined spatial framework persists by usi

block groups[3], credibility and interpretation of patterns is nonetheless improved over mere visual observation as spatial autocorrelation measures such as global and local Moran's *I* coefficients are used to verify the visual impression of clustering. The spatial autocorrelation measures also can give indications of the level of segregation and integration of certain population groups via the multivariate spatial autocorrelation analysis.

The approach uses widely available census data and is easy to replicate for other places. Despite the methodological sophistication, results remain accessible by planners, policy makers and lay people as the statistics are translated into maps. Thus, the approach provides a feasible and easy to use tool for quick assessments of the segregation and residential patterns in urban regions. Planning agencies may use the approach and maps to outline potential zones of conflict and discrimination or evaluate whether certain neighbourhoods face problems in terms of access to public transport, health care or exposure to environmental risks.

11.7 REFERENCES

Anselin, L., 1996, The Moran scatterplot as an ESDA tool to assess local instability in spatial association. In *Spatial Analytical Perspectives on GIS*, edited by Fischer, M. *et al.* (London: Taylor & Francis), pp. 111-125.

Anselin, L., and Bao, S., 1997, Exploratory Spatial Data Analysis: Linking SpaceStat and ArcView. In *Recent Developments in Spatial Analysis*, Chapter 3, edited by Fischer, M. and Getis, A. (Berlin and New York: Springer), pp. 35-59.

Anselin, L., February 25, 2000, SpaceStat^{TM} software for spatial data analysis and related services, Distributed by Biomedware Inc. Copyright Luc Anselin 1999-2000, WEBSITE, Available: http://www.spacestat.com [Accessed May 18, 2000].

Burgess, E.W., 1925, The growth of the City: An Introduction to a Research Project. In *The City,* edited by Park, R. *et al.*, (Chicago: Chicago University Press).

Cliff, A.D., and Ord, J.K., 1981, *Spatial Processes: Models and Applications*, (London: Pion).

Dodge, M. and Kitchin, R., 2000, *Mapping Cyberspace*, (London: Routledge).

Foster, K.A., 1997, *The Political Economy of Special-Purpose Government*, (Washington:Georgetown University Press).

GARM - Geographic Areas Reference Manual, 1994, U.S. Department of Commerce. Economics and Statistics Administration. Bureau of the Census. ONLINE. Available: http://www.census.gov:80/geo/www/garm.html [Accessed May 1999].

Goodchild, M. F., 1986, Spatial Autocorrelation. In *Concepts and Techniques in Modern Geography* (CATMOG), **47**, (Norwich:Geobooks), pp. 1-56.

Gould, P.R., 1970, *Is Statistix inferens* the geographical name for a wild goose? *Economic Geography*, **46**, pp. 439-448.

Hawley, A., 1950, *Human Ecology: A theory of Community Structure*, (New York: Ronald Press).

Heikkila, E., 1996, Are Municipalities Tieboutian clubs? *Regional Science and Urban Economics*, **26**, pp. 203-226.

Johnston, R.J., 1982, *The American Urban Systems: A Geographical Perspective,* (New York: St Martin's Press).

Knox, P., 1994, *Urbanization: An Introduction into Human Geography*, (Englewood Cliffs: Prentice Hall).

---

[3] Dependence on the spatial framework could be weakened by a resampling of values into a grid structure.

Legendre, P., 1993, Spatial Autocorrelation: Trouble or New Paradigm?, *Ecolo* pp. 1659-1673.

Park, R., 1936, Human Ecology, *American Journal of Sociology* **42**(2), pp.1-15.

Park, R., *et al.* (eds.) 1925, *The City,* (Chicago: University of Chicago Press).

Pendall, R., 2000, Local Land Use Regulation and the Chain of Exclusion. *Jour American Planning Association*, **66**(2), pp.125-142.

Perry, C.A., 1929, The Neighborhood Unit, In *Regional Study of New Yo Environs, VII, Neighborhood and Community Planning, Monograph One (N* Regional Plan of New York), p. 88.

Sack, R D., 1986, *Human Territoriality: Its Theory and History,* (New York: C University Press).

Shen, Q., 1994, An Application of GIS to the Measurement of Spatial Autoc *Computers, Environment and Urban Systems*, **18**(3), pp. 167-191.

Suttles, G., 1970, *The Social Construction of Communities,* (Chicago: Uni Chicago Press).

Tiebout, C.M., 1956, A pure theory of local expenditures. *Journal of Political* **64**(5), pp. 416-424.

Tobler, W., 1970, A computer movie simulating urban growth in the Detrc *Economic Geography,* **46** (Supplement), pp. 234-240.

US Bureau of the Census, *1990 Census of Population, Social and Characteristics*, Urbanized Areas, **1-6**, (Washington: Government Printing C

US Bureau of the Census, *1990 Census of Population, Urbanized Ar* Supplementary Report, **1-2**, (Washington: Government Printing Office).

Weiher, G.R., 1991, *The Fractured Metropolis: Political Fragmenta Metropolitan Segregation,* (Albany: SUNY Press).

12

Georeferencing social spatial data and intra-urban property price modelling in a data-poor context: a case study for Shanghai

Fulong Wu

12.1 INTRODUCTION

Despite enormous progress in GIS techniques, the lack of geo-referenced spatial data is still a major constraint to the development of more accurate and in-depth/sophisticated spatial analysis in developing countries. Even when some spatial data are available in theory, the use of geo-referenced data is constrained due to two reasons. First, the digital data may not be available to researchers (especially for those are not affiliated to a local institution) for the sake of confidentiality. For example, in the case of China, population census is not released to the public at the sub-district, *i.e.* Street Office level or lower levels. Even at the sub-district level, the data are not publicly available except for the total population. However, the sub-district is probably equivalent to wards rather than enumeration districts (EDs) in UK, which is still too coarse for the purpose of examining subtle intra-urban spatial variation. Ideally, disaggregated data should be obtained for urban and rural settlements because residential quality varies at the scale of neighbourhoods. In case of China, these units are residents' and villagers' committees.

Second, even when some spatial data are available, it is difficult to directly apply them in spatial analysis because they are collected from different spatial units. For example, population data are usually collected on the basis of subdistricts, while the data of firms are associated with postal codes and land use are interpreted from parcel-based images or maps. The need for cross-referencing different spatial data layers then becomes an issue. In general, the availability of spatial data in developing countries is undesirable. A more spatially sensitive methodology is therefore critical for empirical research in this data-poor context. This project uses disaggregate property data from Shanghai, PRC to study intra-urban spatial variation of property prices. This is achieved through geo-referencing the address information and interpolation of residual surfaces, which highlights the way forward in developing high-resolution spatial data for decisic in developing countries.

12.2 GEO-REFERENCING PROPERTY DATA

From the social scientist's perspective, the key issue of spatial representatic referencing socio-economic phenomena (Martin, 1999), through for example, a the property address with the national grid reference (Martin and Higgs, 1996). there has been a massive improvement in the resolution and quality of geo-re socio-economic data (Martin, 1999). The resolution for population data increase km grid squares in 1970s, to digitised ED boundaries including about 400 pers early 1990s, and now to 0.1 m Address-Point information. Thanks to the spatia functions, GIS is naturally the ideal tool for real estate valuation. Higgs *et al.* (1 the distribution of properties in council tax bands, based on a survey of proper Using GIS analysis techniques, Longley *et al.* (1994) assess the accuracy of c banding in Cardiff. GIS-based property information system has been deve integrating various sources of spatial data (Wyatt, 1997). The digital geo-refere map product such as ADDRESS-POINT and Land-Line.Plus produced by the Survey (OS) together with population census at the Enumeration District (ED) I made it possible to develop property price models (Lake, *et al.*, 1999, 200C 1999; 2000). Through GIS operations and processing functionality, a wide neighbourhood attributes describing the character of a property's surroundin; inferred. For example, based on DEM, viewsheds can be calculated to assess impact of amenity and nuisance land uses (Lake, *et al.*, 1998).

In contrast, the spatial analysis of property price variation is difficult in d countries because of the lack of high resolution spatial data. Efforts have been using indirect measurements from aerial photographs and remote sensing in study the settlement structures (*e.g.* de Bruijin, 1992) but these are only within of photogrammetric engineering and land cover studies. In the case of Chine there have been a few attempts to overcome the data constraint by deriving in directly from aerial photographs and satellite imagery (such as Landsat TM ima and Yeh, 1997; 1999; Yeh and Li, 1997; Li and Yeh, 2000). Ortho-adjus photographs so far provide the most accurate information on land uses. On a scale, roads and large buildings are identifiable. Landsat TM at a resolution metres can be used to identify the conversions from the rural to urban lan particularly useful in finding construction sites because of significant chang spectral characteristics of land surface. Through making reference of factory and centroid of urban sub-districts, the population and firm potentiality surf derived through spatial interaction formula. Based on these variables, intra-u use probability models of Chinese cities have been developed (Wu, 1999). Hov potential of the rich socio-economic information source has not been fully Recently, intrametropolitan location of foreign investment firms has been mo geo-referencing their post-codes (Wu, 2000). Registered on the same co-ordinat this allows the cross-reference of foreign investment with the land use in derived from aerial photographs and Landsat TM imagery. Previous studies s geo-referencing of socio-economic data could provide high potential for develor urban models.

12.3 INTRA-URBAN PROPERTY PRICE MODELLING

Traditionally, urban property price has been modelled under the hedonic price framework developed by Rosen (1974), in which the property is regarded as a bundled commodity consumed in a competitive market. The complication of housing price modelling is that the attributes are not limited to the product itself because of the existence of externalities. This has led to the inclusion of attributes at different levels of resolution which can be conceptually divided into structural characteristics, accessibility characteristics, and neighbourhood characteristics (Olmo, 1995). The latter is often a determinant of the dynamics of local housing market. While empirical studies in the 1970s used very simplistic measures of locational features (such as the distance to CBD), the advance of spatial database technology is driving recent research towards more sophisticated way of capturing the characteristics of the environment in which the property is located. For example, Lake *et al.* (1998) used GIS viewshed analysis to include the visual impact of various land uses on property prices. Orford (1999) developed explicit measures of the attributes at the levels of property, street, and community. GIS is naturally an ideal useful tool to derive various 'spatial' attributes through spatial operations (such as distance proximity, buffer, overlay, network analysis, visibility analysis). By cross-referencing different layers, new attributes can be generated, for example from the Census, unit postcode areas, and other utility databases. In a sense, the utilisation of GIS can help develop more context-sensitive property price modelling (Orford, 2000). The list of attributes that can be generated from GIS operations seems endless, although this can be a problem when determining which ones are significant.

The rationale behind using various locational variables is that, in order to estimate a proper contribution of the attributes to property prices, one has to consider *local* housing market dynamics. Orford (2000, p.1644) argues that, 'if the hedonic house price function is to generate estimates that properly reflect the implicit price of attributes, the model specification must capture sufficiently the dynamics of the local housing market. In particular, the specification must reflect the variations in supply and demand of housing attributes and also the ad hoc nature of the valuation process'. The problem underlying the use of ordinary least-squares (OLS) regression in the estimation of the hedonic property price model is the existence of heteroscedasticity and spatial autocorrelation in the data, which violates the assumptions of independent, identically distributed errors. In other words, the contribution of a factor (such as the size of the property) to the price may vary spatially – in accordance with different dynamics of a local housing market. Based on this understanding, Orford (2000) proposed a multilevel modelling approach. The specification of multilevel models includes the attributes at the property, community (wards) and EDs levels. The model therefore allows more explicit considerations of spatial variation and interaction among different levels of attributes. In particular, it is important to justify that the existing boundaries of wards and of EDs delineate the sub-markets of housing. In other words, the development of a property price model requires a much deeper understanding of a particular housing market which is different from place to place.

12.4 METHODOLOGY

The classic specification of the hedonic price model includes the attributes of property structure and locational characteristics (Can, 1990; Olmo, 1995):

$$p_i = \alpha + \sum_{k=1}^{K} \beta_k S_{ik} + \sum_{j=1}^{J} \delta_j L_{ij} + \varepsilon_i,$$

where i = 1 to N, N is number of properties, p_i = the price of ith property attribute of property structure; L_j = jth locational characteristics of the ith prope and δ_j = regression coefficients, ε_i = disturbance term that is only related property.

This hedonic approach, however, does not consider the interaction property attributes and neighbourhood characteristics. Oxford (2000) used a modelling approach to capture the influence of local community on housi However, it is difficult to apply this approach to Shanghai because the def 'community' is unclear. The urban districts designated by Shanghai government for the administrative purpose are larger than the wards defined in Cardiff, Orford (2000, p. 1651-1652) finds that the 'communities' used by council are defined by major thoroughfares, rivers and railway line, and communities are often used by estate agents as a basis for defining neighbourhoods. In contrast, urban districts in Shanghai range from 0.2 (Huangpu District) to 1 million (Yangpu District) (SSB, 1998). It is unlikel districts of this size could form a unitary housing market. Below the urban di sub-district offices. In 1997, there are totally 99 sub-district offices in the m area. However, the sub-district offices are traditionally not a substantial unit management because of centralised administration. Data at the sub-urban distr are sparse and difficult to collect. The central point is that it would not be very include *ad hoc* defined administrative units into property price modelling needed, however, is to show the city-wide distribution of the deviation of prop at the resolution of individual properties.

In this study, we define the structure attributes as those that are only rela property itself. All locational features are not included, because these attr difficult to measure and the spatial structure of these attributes are und regression thus gives an error term that might be spatially autocorrelated. As a regression cannot be interpreted in a way it is normally used. But the purp examine the residuals once the structure variables are controlled.

$$p_i = \alpha + \sum_{k=1}^{K} \beta_k S_{ik} + \varepsilon_{xyi},$$

where, x,y = location co-ordinates of the property i.

The method is therefore to separate the environmental influence fro property structure, while the interaction of environmental and structural att ignored. The underlying assumption is that the adjustment of the property pr not be dependent on the property but rather on the environmental amenity. H needed, it would be possible to classify residuals according to the type of pro generates the residual. As a result, there will be a series of residual surface ac property types. However, certain property types may be located in part of the therefore it is likely that the surface will only cover part of the city.

The influence of spatial factors can be reflected through the deviation o price from the metropolitan-wide 'standard' price that based on the structure of

(Figure 12.1). The deviation of the observed price from the price predicted through the attributes is the 'residual' of regression, which is composed of the 'unknown' disturbance in property transaction and a locational premium (positive or negative). There is no reason to believe that the disturbance in property transaction should present a systematic spatial variation.

In order to examine the spatial distribution of 'residuals', the residuals are interpolated from the sample points into a surface. This interpolation of residual surface is similar to the development of Digital Terrain Model (DTM). The residuals can be interpolated because they reflect the contribution from the environment, which is seen as continuous over space. Certain spatial barriers such as the river and major railway lines may separate communities into distinct sub-markets. However, this is not modelled in the current study.

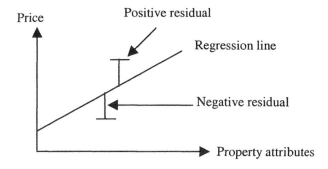

Figure 12.1 The regression line of property price and positive and negative residuals.

There are a variety of interpolation methods (Burrough and McDonnell, 1998). The most common one is the inverse distance weighted (IDW) interpolation method, which is used to generate the residual surface from the point coverage of the regression model. The formula for calculating the output cell value is:

$$c_{x1y1} = \frac{\sum \varepsilon_{xyi} d[(x, y), (x1, y1)]^{-s}}{\sum d[(x, y), (x1, y1)]^{-s}}, \qquad (12.3)$$

where $x1$, $y1$ = co-ordinates of the centre of the cell; c = the value of interpolated residual; $d[(x,y), (x1,y1)]$ = the distance between the centre of the cell to the point location; s = distance decay parameter.

The IDW interpolation can also specify a neighbourhood range. This assumes that the contribution of the samples outside this range should be neglected. We use different ranges to examine the contribution of environmental factors at the local (400 m) and community (1,000 m) resolution. When the distance decay parameter is specified as zero, the distance decay within the neighbourhood is not considered. In essence, the cell value produced by the interpolation becomes the average of the residuals within the moving kernel. This is conceptually plausible, as this will in fact 'smooth' out the random effect of property transaction and leave the 'regional' variation (Figure 12.2).

The fixed neighbourhood range only generates the grid cells where at sample point exists within the kernel. From the property price modelling point means we cannot predict the property value where no sample is available. Th sampling locations are mainly located in the suburbs of the city. However, it that because of the large size of land plot and relatively homogenous price vari possible to use the sample from a longer distance than in the inner cities. Alt the minimum number of points can be specified to interpolate the residual surf the specified minimum points cannot be found in the kernel, the search radius (to the neighbourhood range) is enlarged to ensure that the minimum number o used. If the interpolation of fixed neighbourhood range is applied to the property price, this gives the average property price within the moving kernel.

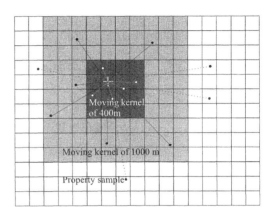

Figure 12.2 The interpolation of residual surface from the property sample by specifying a moving kernel of 400 m or 1,000 m influence.

12.5 CASE STUDY OF SHANGHAI

12.5.1 Data Source

This project aims to find the spatial differentiation of Chinese cities in the towards a more-market oriented economy. Because of the lack of high-population data in China, it is extremely difficult to study intra-urba differentiation. However, the information of property price can be used to ex urban spatial changes which are becoming more differentiated in the newly e real estate markets.

The main information comes from the *Shanghai Housing Market* releas Shanghai Real Estate Exchange Centre (SREEC). The data consist of a total records of the residential properties for sale in August of 2000. The size of t was then reduced to 1,604, using stratified sampling according the total n properties available for sale in each district. This was due to the heavy worklo referencing their locations. The distribution of properties in the urban d Shanghai is shown in Table 12.1. The location of each property was geo-refe hand through its address because no digital product is available. This is a ve. and time-consuming process.

12.5.2 Map Registration and Transformation

The basic location information is based on *Shanghai Atlas* (1997), which is in turn mainly derived from the 1:10,000 topographic map produced by Shanghai Survey Academy in 1996. The maps from the Atlas were scanned, registered with control points and then rectified. The Atlas also provides an address index, which allows a quick finding of a street through its name, and some key road numbers along each street. Therefore, the precise address location is unknown and has to be inferred from nearby key addresses (Figure 12.3).

Table 12.1 The distribution of the sample of properties in Shanghai, 2000.

District	No. of properties	Percentage in total
Baoshan	52	3.4
Changning	159	10.3
Hongkou	114	7.4
Huangpu	19	1.2
Jiading	14	0.9
Jing'an	110	7.1
Luwan	71	4.6
Minhang	138	9.0
Nanshi	30	1.9
Pudong	174	11.3
Putuo	273	17.7
Xuhui	154	10.0
Yangpu	161	10.4
Zhabei	72	4.7

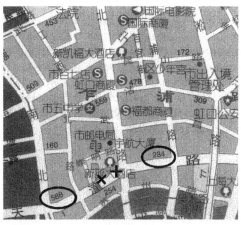

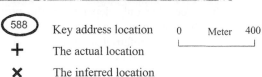

Figure 12.3 Identification of the location from the street address.

This process is subject to error and distortion. In order to assess the degre
some tests are carried out to infer known addresses through this method.
researcher was given an address and asked to infer 'blindly' its possible loca
the nearby key addresses. Then, the actual location is revealed and the d
deviance was recorded. The accuracy of inferring a location by the post-cod
sub-districts is also assessed. The size of post-code areas increases from the c
to rural areas and therefore the error produced will depends on the distribut
location. But from the test the manual reference of street address is within the a
50 to 100 metres. In some extreme cases, the variation can be as large as 5(
Compared with the geo-referencing product such as ADDRESS-POINT, the re
accurate but much better than other possible solutions in a data-poor situation.

The study area consists of 14 districts. Each map sheet in the Atlas c
urban district. The co-ordinates derived from the referencing address on each
thus are measured in map projection. These co-ordinates need to be transforr
single common co-ordinate system in order to view the metropolitan wide distr
property price. The common co-ordinate system is referred through the large-
provided by the Shanghai Survey Academy. Because the map sheets are der
ortho aerial photographs, 'rubber sheeting' is not required. The registration of r
only requires the transformation of map units into the units of the common c
system. For each map sheet, the following steps are applied:

1. establish the tics (control points) on the basis of obvious land marks suc
   intersection and bridges;

2. measure the co-ordinates of these tics in map unit as well as real-world un

3. create a new coverage using the tics with their real-world co-ordination;

4. transform the co-ordinates of map projection into the newly created cover
   measured in the co-ordinates of all districts.

Finally, all transformed coverages are merged into a single coverage o
location. On the basis of property-ids, the property attributes are then jo
locations through point-attribute table (PAT) to form the database of propert
are then used for property price modelling. Figure 12.4 shows the locati
properties. In Pudong, a new district created in 1990, some properties are devel
the newly built roads, and so it is not possible to refer the location thro
addresses. Consequently, these properties are dropped out from the analysi
number of properties in the database is reduced to 1,431.

12.5.3 Price Modelling

In this study, the use of the hedonic price approach is not to provide the re
between the property price and the property attributes. Rather, the regression
control the effect that only depends on the property itself, thus separating t
effect through the examination of the residuals. The deviation of the observed p
the predicted price is referred to as a residual. The residual can be due to '
disturbance purely related to the property itself and location premium. The forn
not present a systematic spatial variation and the latter can be visualised to
intra-urban spatial changes because the properties are 'geo-referenced'. T
variation can be imagined as a rough surface. This is because the unknown ar

noise also contributes to the shape of the surface. The point-based observation is then interpolated into a surface. The model, to our current knowledge, is the first of its kind to be developed in a data-poor context. While the hedonic price approach has been widely applied, only through geo-referencing the observed data is it possible to visualise the spatial variation of location premium.

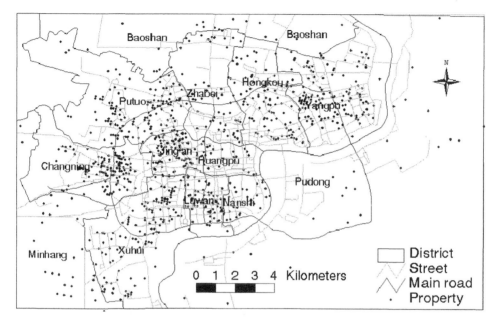

Figure 12.4 The distribution of sampled properties in Shanghai, 2000.

The regression uses the physical and property right attributes of housing. The first category includes the number of rooms (ROOM), the number of receptions (REPT) and the second category includes two dummy variables SRT and PRT, which represents respectively the full property right and partial property right in a 'commodity housing' market. The full property right means the property can be sold openly in the housing market without any restriction, while the partial right housing means the use right can be sold but the owner does not have the full right of the land premium associated with the house. If both dummy variables are zero, the property belongs to the 'public housing', which cannot be sold in housing market. These properties can be exchanged if certain compensation is paid to their occupants.

Table 12.2 suggests that the physical attributes of housing are good predicator of housing price with R^2 of 0.619, which the property right alone only partly explains the housing price, with a low R^2 of 0.052. The full attribute model gives the R^2 of 0.619 and is a controlled prediction. The residual of the model is calculated accordingly.

12.5.4 Residual and Property Price Surface

The residual interpolation was undertaken using ESRI Arc/Info GRID module. The output table of statistical analysis in SPSS is joined with the PAT of the processed property sample coverage. The residual value is used as the 'spot' item for the interpolation command. Two neighbourhood ranges are used to represent price at different scales. The command used for generating the surface is:

```
Residual_surface = pointinterp( housing_sampling_cover, residual_va
cell_size, idw, decay_parameter, plateau_parameter, the
neighbourhood_shape, size of neighbourhood)
```

Table 12.2 Multiple regression of housing price (dependent variable: housing price in 10,00

	Physical attributes (the whole city)		Property right (the whole city)		Full attribute (the whole ci
	B	t	B	t	B
Constant	-13.285	-17.960**	10.827	9.067**	-12.508
ROOM (No. of rooms)	16.460	46.595**			16.525
REPT (No. of reception)	3.887	6.214**			4.255
SRT (Dummy variable, with reselling right)			6.360	4.083**	-1.470
PRT (Dummy variable, with property right)			12.974	9.046**	-1.324
F ratio		1248.571		43.322	
R²		0.619		0.052	
No. of cases		1,541		1,541	

Notes: B = coefficient of regression; t = t-values; ** significant at the 0.01 level and
* significant at the 0.05 level.

The cell size of the surface is 100 m, the decay parameter and plateau (within the distance the weighting of points is constant) are set to zero. Th neighbourhood is set respectively as 400 m and 1,000 m. Figure 12.5 and 12.6 generated residual surfaces.

The surfaces clearly show that the areas in the Xuhui District and part district, where the previous French International Settlement was located, ha average positive residuals. The residential quality of these areas is higher tha places as Yangpu and Zhabei District where there is a high proportion of indus A second clustering of positive residual is near North Shichuan Road in the District. For the new urban district, Pudong District, because there are not enou points, the residual surface covers only some locations and does not show th variation.

To measure the spatial correlation, the *Moran I* is used as a spatial indicator reflecting the degree of spatial autocorrelation (Goodchild, 1986). The is used to reveal the pattern of clustering of the same value at adjacent cells. Th provided by the ARC/INFO *Moran I* function does not describe the clustering a of spatial object. The absolute concentration of value within a kernel generates close to unity, while a more even distribution than can be expected by chan value below zero.

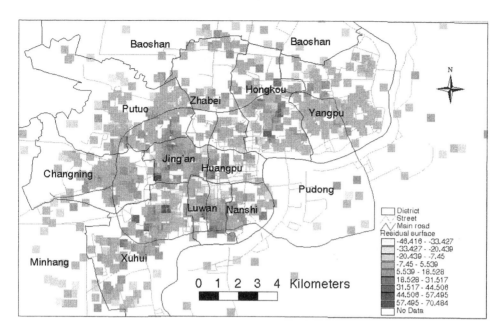

Figure 12.5 The residual surface at the resolution of 400 m neighbourhood range (unit: 10,000 Yuan).

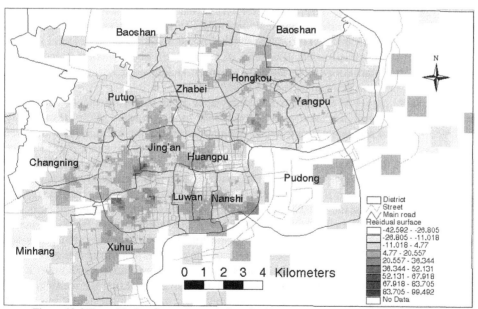

Figure 12.6 The residual surface at the resolution of 1,000 m neighbourhood range (Unit: 10,000 Yuan).

Although the absolute value of Moran I may not correspond to a fixed scale of spatial clustering, the indicator can be used to assess the degree of spatial autocorrelation. Table 12.3 shows the Moran I for the four surfaces.

Table 12.3 The Moran I measurement of the interpolated residual and price grids.

Neighbourhood Range	Residual surface	Price Surface
400 m	0.74038	0.7350
1,000 m	0.8003	0.8901

For the interpolation of the residual surface using the minimum point radius, the command used is:

```
Residual_surface = idw( housing_sampling_cover, residual_value,
barrier_cover, decay_parameter, radius, radius_parameter, minimum_r
cell_size)
```

The decay parameter is set to 2; the radius parameter sets the initial s search radius (the default is 5 times the cell_size); minimum points are set t command generates the interpolated surface, which covers the area where tl points are not available.

The surface, presented in Figure 12.7, clearly reveals the clustering of th contribution of environment factors to housing prices. The darker areas hav residuals, which means the environment enhances the property price; the lig (including Huangpu) have the negative residuals, which means the areas attractive in terms of property value. The positive area are generally distribu with the circular road, clustering at some particular locations with high ɪ quality; these areas to a certain extent conform to the legacy of high-class ɪ areas before the establishment of socialism in 1949. The negative areas are dist the inner areas (such as the Huangpu) and industrial areas (such as Putuc Yangpu), and peripheral districts (Baoshan and Minhang) where infrastructu developed. The situation in the Pudong new district is less clear because th interpolated over a longer distance (than 400 m).

The same approach is applied to the property price to derive the aver within a moving kernel. The size of the kernel is set to 400 m and 1,000 m re for neighbourhood and community levels. Figure 12.8 shows the average price of standard deviation from the mean. The darker values indicates the average pɪ the mean, while the lighter shades reflects the opposite. It can be seen that prop vary within each district. In general, the property price in peripheral loca industrial areas such as the Baoshan, Putuo, and Yangpu District is lower than ɪ Luwan, and Xuhui District. In the inner city (the Huangpu District) property pr because of the high proportion of old 'alley housing' (terrace housing). For th 1,000 m kernel, the concentration of high property price becomes even mor (Figure 12.9). Because of internal price variation, the average price in Hongkc is lower than some places in the Jing'an, Xuhui, Luwan and Changning Distr compare the prices of Zhabei and Hongkou District, it is obvious that the mea price is different. This is consistent with the understanding of the different these two districts.

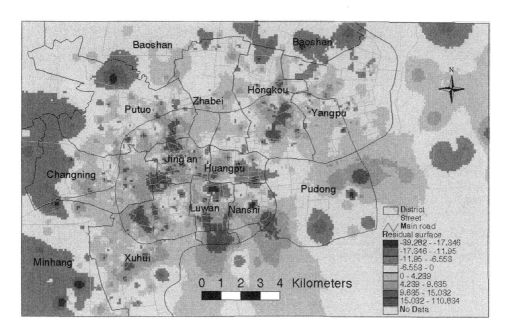

Figure 12.7 The residual surface interpolated using the varying radius with a minimum number of eight sample property locations (unit: 10,000 Yuan).

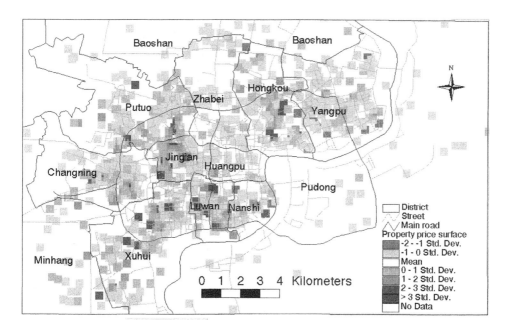

Figure 12.8 The average property price surface with the 400 m neighbourhood range. (Unit: 10,000 Yuan).

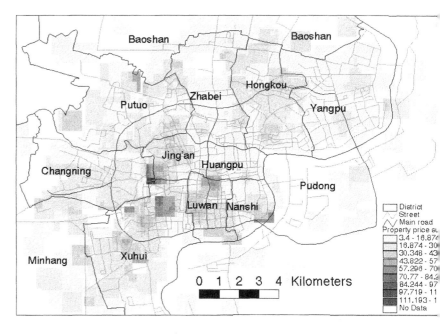

Figure 12.9 The average property price surface with the 1,000 m neighbourhood range (unit: 10,0

12.6 CONCLUSION

This work is the first of its kind to explore social-spatial differentiation in the co
developing country where socio-economic data at a high spatial resolution a
difficult to obtain. Through geo-referencing the address information, it is p
construct a database of the distribution of activities at the highest possib
resolution. Under the data-poor context, small-area data (such as these based
tract and enumeration district) will likely take many years to develop, while ra
and economic changes require that up-to-date information be provided t
decision-making. However, to speed up data capturing, it is critical to automat
referencing process. Manual processing is still too time consuming to provide
solution. A possible way forward is to register all the addresses and use
matching' to automate the process. This may prove to be very costly in a
context because this requires a large amount of financial support. Alternati
possible to use 'key addresses' as shown in this study to infer the location c
between them. Although this is subject to error and distortion, this approac.
practical and low-cost in the development of disaggregated spatial data. Onc
addresses are digitized, the new address can be created using 'interpolation'. I
new address between the key addresses can be created during the data autom
thus do not need to be stored.

Regarding the property price modelling, in the data poor context, the
should be the visualisation of the distribution of positive and negative envi
contribution to property prices. The study shows that by geo-referencing the l
properties and the development of a spatial property database, it is possible to q
the interpolation method to display the spatial variation of property price di

The results of the study reveal that property market in Shanghai shows the feature of spatial differentiation. Future research should use various additional sources of data about the distribution of facilities, services, population, and infrastructure to study the reasons for spatial differentiation of the housing market. Methodologically, the automation of geo-referencing socio-economic data will be important, as this will open up a rich source of information in addition to government statistics.

12.7 ACKNOWLEDGEMENT

The study is supported by the Nuffield Foundation project (SGS/LB/0488).

12.8 REFERENCES

Can, A., 1990, The measurement of neighbourhood dynamics in urban house prices. *Economic Geography*, **66**, pp. 254-272.

de Bruijin, C.A., 1991, Spatial factors in urban growth: towards GIS models. *ITC Journal*, **4**, pp. 221-231.

Goodchild, M.F., 1986, *Spatial Autocorrelation: Concepts and Techniques in Modern Geography 47*, (Norwich: Geo Books).

Higgs, G., Longley, P. and Martin, D., 1992, Analysing the spatial implications of the Council Tax: a GIS approach. *Mapping awareness and GIS in Europe*, **6**, pp. 42-46.

Lake, I.R., Lovett, A.A., Bateman, I.J. and Day, B., 2000, Using GIS and large-scale digital data to implement hedonic pricing studies. *International Journal of Geographical Information Sciences*, **14**, pp. 521-541.

Lake, I.R., Lovett, A.A., Bateman, I.J. and Langford, I.H., 1998, Modelling environmental influences on property prices in an urban environment. *Computers, Environment and Urban Systems*, **22**, pp. 121-136.

Li, X. and Yeh, A.G.O., 2000, Modelling sustainable urban development by the integration of constrained cellular automata and GIS. *International Journal of Geographical Information Sciences*, **14**, pp. 131-152.

Longley, P., Higgs, G. and Martin, D., 1994, The predictive use of GIS to model property valuations. *International Journal of Geographical Information Systems*, **8**, pp. 217-235.

Martin, D., 1999, Spatial representation: the social scientist's perspective. In *Geographical Information Systems: Principles, Techniques, Management and Applications*, edited by Longley, P.A., Goodchild, M.F., Maguire, D.J. and Rhind, D.W. (Chichester: Wiley), pp. 71-80.

Martin, D. and Higgs, G., 1996, Georeferencing people and places: a comparison of detailed datasets. In *Innovations in GIS 3*, edited by Parker, D. (London: Taylor and Francis), pp. 37-47.

Olmo, J.C., 1995, Spatial estimation of housing prices and locational rents. *Urban Studies*, **32**, pp. 1331-1344.

Orford, S., 1999, *Valuing the Built Environment: GIS and House Price Analysis*, (Aldershot: Ashgate).

Orford, S., 2000, Modelling spatial structures in local housing market dynamics: a multilevel perspective. *Urban Studies*, **37**, pp. 1643-1671.

Rosen, S., 1974, Hedonic prices and implicit markets: product differentiatio competitions. *Journal of Political Economy*, **72**, pp. 34-55.

State Statistical Bureau (SSB), 1998, *Shanghai Statistical Yearbook 1998*, China Statistical Press).

Wu, F., 1999, Polycentric urban development and land use change in a tr economy: the case of Guangzhou. *Environment and Planning A*, **30**, pp.107;

Wu, F., 2000, Modelling intrametropolitan location of foreign investment i Chinese city. *Urban Studies*, **37**, pp. 2441-2464.

Wu, F. and Yeh, A.G.O., 1997, Changing spatial distribution and determinan development in China's transition to a market economy: the case of Gr *Urban Studies*, **34**, pp. 1851-1879.

Wu, F. and Yeh, A.G.O., 1999, Urban spatial structure in a transitional econom of Guangzhou, China. *Journal of the American Planning Association*, **65** 394.

Wyatt, P. J., 1997, The development of a GIS-based property information syste estate valuation. *International Journal of Geographical Information Scienc* 435-450.

Yeh, A.G.O. and Li, X., 1997, An integrated remote sensing and GIS appro; monitoring and evaluation of rapid urban growth for sustainable developm Pearl River Delta. *International Planning Studies*, **2**, pp. 193-210.

PART IV

GIS and Rural Applications

13

Accessibility to GP surgeries in South Norfolk: a GIS-based assessment of the changing situation 1997-2000

Andrew A. Lovett, Gisela Sünnenberg and Robin M. Haynes

13.1 INTRODUCTION

A guiding principle of the National Health Service in the UK is the ideal of equal access to health services for those in equal need. Health services, however, are inevitably concentrated in certain places, and are consequently more accessible to nearby residents than people living further away. Inequality in accessibility due to distance is, of course, only one element of the problem of equal access to health services. Access to services can be represented as a continuum, across which many social, economic and cultural factors may contribute (Ricketts and Savitz, 1994), but the difficulty of overcoming distance tends to be a dominant factor in rural regions (Joseph and Bantock, 1982). Poor physical accessibility is known to reduce the use of services, and may lead to poorer health outcomes (Joseph and Phillips, 1984; Haynes, 1986; Jones and Bentham, 1997). Low utilization of primary care services is of particular concern because of the gateway role of general practitioners (GPs) in terms of referral to hospitals. There is a considerable body of research on variations in physical accessibility within rural areas of the UK (*e.g.* Moseley, 1977; Higgs and White, 2000), and the problem is still an important concern for several government agencies (Department of Environment, Transport and the Regions, 2000; Countryside Agency, 2001).

Developments in Geographical Information Systems (GIS) and digital map databases have made it possible to calculate measures of physical accessibility such as travel time in a more automated and sophisticated manner than was previously practical (*e.g.* Higgs and White, 1997; Naude *et al.*, 1999). Another innovation has been the use of GIS to assess accessibility by public transport, taking into account the spatial distribution of bus routes and frequency of services (O'Sullivan *et al.*, 2000; Higgs and White, 2000). Both these types of GIS techniques were used in a study of access to primary health care services in East Anglia during Autumn 1997 (Lovett *et al.*, 2000). The results of this analysis included evidence of an inverse care law; in the sense that the rural populations with the poorest accessibility were also more deprived and with evidence of higher health needs than other populations.

Since 1997 there have been considerable changes to public transport s
rural areas of the UK. In the Spring 1998 Budget the government announced
additional funds to support rural bus services and later that year the concep
Transport Partnerships was introduced to encourage local community transport
Both these developments have resulted in new services for rural populations i
such as East Anglia, but there have also been instances where services
withdrawn or altered only a few months after being introduced. It was therefo
appropriate to assess what had been achieved and examine the extent to which
in accessibility still remained. To achieve this objective, an investigation of
accessibility to GP surgeries between Autumn 1997 and Autumn 2000 was unde
the district of South Norfolk, working in co-operation with the local authority
Norfolk Primary Care Group (PCG).

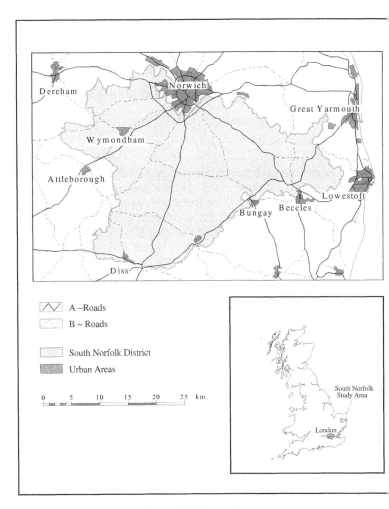

Figure 13.1 The study area.

South Norfolk covers an area stretching from several suburbs of Norwich to the county border with Suffolk (see Figure 13.1). It includes several market towns and had a population of approximately 104,000 in the 1991 Census. South Norfolk was placed in the 'Prospering Areas' cluster in the Office for National Statistics (ONS) classification of local authorities (Wallace and Denham, 1996), but in the previous East Anglian study (Lovett *et al.*, 2000) it was found to include several groups of parishes with the poorest accessibility to primary health care services. The district was therefore considered to encompass a good range of accessibility conditions, although it is fair to note that even the most rural resident would not be more than 20 km from a market town and so would not be classed as remote in comparison to some regions of the UK.

13.2 DATA

13.2.1 Primary Care Facility Locations

Details of GP surgeries were supplied by the health authorities and PCGs covering South Norfolk and the surrounding area. Information was obtained on main and branch surgeries existing in Autumn 1997 and Autumn 2000. Outlying consultation facilities that were typically open for only a few hours each week were not included in the analysis. All the surgery addresses were converted to 100 m resolution grid references via the Postcode to Enumeration District Directory (Martin, 1992). Between the two dates there were only three short distance moves by practices to new premises so the overall pattern of facility provision did not alter appreciably.

13.2.2 GP Patient Registers

Information on the resident population of South Norfolk in Autumn 1997 and Autumn 2000 was derived from GP patient registers. This source was used because it supplied population estimates that were as contemporaneous as possible with the other data and provided a level of geographical detail that was appreciable better than alternatives such as ONS mid-year population estimates or 1991 Census products. The geographical resolution of population data is an important consideration in accessibility research, because analyses based on centroids of areas such as enumeration districts or parishes inevitably tend to over-concentrate residents in locations where services are more likely to be present (Lovett *et al.*, 2000).

Downloads of selected fields in the patient registers for South Norfolk in Autumn 1997 and 2000 were obtained from the health authorities and PCGs. All the records were anonymous and contained details of number of residents for each unit postcode and GP practice code combination. The unit postcodes were converted to 100 m resolution grid references using the Postcode to Enumeration District Directory and the positions were checked using point-in-polygon techniques with enumeration district boundaries (Gatrell *et al.*, 1991) and distances from ED centroids. Corrections were made using Internet resources such as http://www.streetmap.co.uk where necessary.

Some doubts have been expressed about the reliability of GP patient registers as a source for local population estimates, but previous research (Haynes *et al.*, 1995; Lovett *et al.*, 1998) has concluded that those in East Anglia are of increasingly high quality. Table 13.1 compares the two patient register totals for South Norfolk with recent ONS mid-year estimates. The patient register figures are clearly slightly higher than those from ONS, and this may be due to list inflation arising from administrative delays in removing outdated records (Simpson, 1998). It is worth noting, however, that when the Autumn

2000 patient register figures were compared with parish mid-1999 estimates pr
Norfolk County Council (2000), the largest differences were all concentrated ii
where there is currently considerable house building activity (*e.g.* Wymondham
and the variation for some other urban areas (*e.g.* Costessey on the edge of Nor
much less pronounced. This suggests that the patient register data may be reflec
movements into new residential developments in a quicker manner than is poss
mid-year estimation methodology. Overall, therefore, it was concluded that tl
data were sufficiently reliable to form the basis of the accessibility assessment.

Table 13.1 Population estimates for South Norfolk.

Year	ONS Mid-Year Estimate	GP Patient Register Estimate (Autumn)
1997	106,600	108,314
1998	107,700	-
1999	109,100	-
2000	110,400	113,380

13.2.3 Area Boundaries

For some elements of the research it was necessary to use classifications of ge
areas. Discussions with local authority staff revealed that many aspects of public
provision used parishes, rather than census wards, as a basic spatial unit for pla
management purposes. Parishes also had the advantage of providing greater ge
disaggregation than wards and so were selected as the unit of analysis.
boundaries for the parishes in South Norfolk were constructed from map
available via the MIMAS service at the University of Manchester. There are
118 parishes in South Norfolk, but that around Wymondham (NG117) is far
both areal extent and population) than any of the others. For the purposes of inv
accessibility issues it was thought best to subdivide this parish and treat the sou
as a separate zone. In this chapter, therefore, the term parishes is used to refer
119 areas. All the GP patient register postcodes were matched to parishes via a
polygon procedure and these assignments were checked using the enumeration
parish lookup table in the Area Master File available from MIMAS.

13.2.4 Road Network and Travel Speeds

The road network in the study area is characterised by main roads radiating
Norwich (see Figure 13.1). Details of the road network were taken from the
scale Bartholomew digital map database for Great Britain. The data included
different road categories (*e.g.* A-road dual carriageway, B-road single carriage
average car speeds for each road class were used to calculate how long it wou
drive along any particular segment. The road speeds were based upon nationa
estimates published by the Department of Environment, Transport and the Regi
1993), but reduced slightly to take account of local conditions. These modifi
had previously proved realistic in other research (Bateman *et al.*, 1999).

13.2.5 Bus Services

Information on bus services across South Norfolk was obtained by examining published timetables and route maps supplied by the Norfolk Bus Information Centre. Where necessary, checks were made with local authority staff to produce lists of services as of Autumn 1997 and 2000. For the purposes of assessing accessibility to GP surgeries, particular attention was focused on routes where there was at least one daytime return journey every weekday (*i.e.* a viable level of provision for many activities), but overall the bus services were divided into the following five categories:

1. Routes where there were four or more buses in each direction between approximately 8am and 6pm every weekday and a return journey was possible within three hours.

2. Routes meeting similar criteria to those above, but with only one to three buses in each direction. Services restricted to school terms were not included in this category.

3. Routes providing a journey-to-work service every weekday. These were defined as situations where the outward journey was early in the morning and the return trip did not occur until the evening (*e.g.* someone reliant on such a service would have to spend the whole day at the destination).

4. Routes with a daytime return service (as defined above) on two to four weekdays.

5. Routes where there was a daytime return service on one weekday.

Bus routes were defined on the digital road network using editing facilities in the Arc/Info GIS software. Figure 13.2 depicts the pattern of services in Autumn 2000, individual sections of road being symbolised according to the combinations of provision along them. All of the South Norfolk parishes were also classified into six categories of bus service provision for both Autumn 1997 and 2000. The classifications were based on the minimum level of service available to the majority of the population in each parish, five of the categories corresponding to those for the routes listed above and the sixth representing no bus service. Such an approach is helpful for assessing change, but it needs to be recognised that there were a number of cases where one part of a parish was rather better served than the rest.

13.2.6 Community Transport Provision

Details of community car and dial-a-ride schemes operating within South Norfolk were obtained from service providers and local authority staff. Community car schemes typically involve volunteer drivers using their own cars to provide door-to-door journeys for people without transport, while dial-a-ride services often use minibuses or taxi-style vehicles. The extent of community transport in the district was known to have increased considerably since 1997 and the research focused only on the provision in Autumn 2000. Most of the schemes had defined catchment areas and information on these was used to classify parishes according to the services available. One scheme (CarLink) run by Norwich and Norfolk Voluntary Services operated over the whole of the district and supplied data on the locations of their volunteer drivers as of October 2000. This information was subsequently used to identify gaps in driver availability.

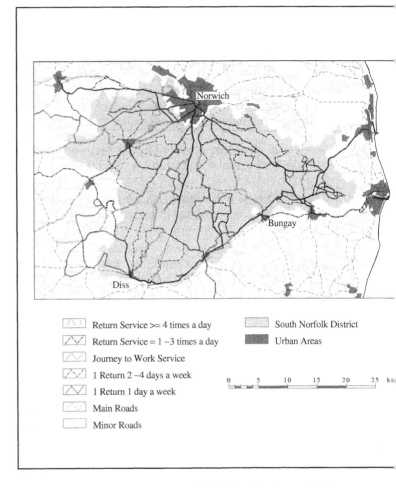

Figure 13.2 Bus routes in South Norfolk, Autumn 2000.

13.3 METHODS

13.3.1 Calculating Journey Times

To calculate travel times along roads for each year it was necessary to first assi
surgeries to their nearest nodes on the network. An allocate operation
implemented in Arc/Info to identify the shortest travel time to a GP surgery
node on the road network. Travel times from patient register postcodes were
using the following procedure. The first step was to identify the closest node o
network for each postcode and then the travel time from that node across
network to the nearest GP surgery. Next, a straight-line distance from the n
postcode was determined and subsequently converted to a time assuming a veh
of 10 miles per hour. This stage was considered necessary because many postc
several hundred metres away from the nearest road on the network (the database
most small residential streets). Adding the two times produced an overall val

postcode and, though the use of straight lines to nodes inevitably introduces a small element of approximation into the calculations, this approach was found to be the most robust option available given the restrictions of the available road network.

Visual representations of travel time by car to GP surgeries in 1997 and 2000 were generated by defining contours around the estimated time values for the road network nodes. This was done within Arc/Info by first undertaking a Delaunay triangulation of the nodes and then creating a continuous surface known as a Triangulated Irregular Network (TIN) (see Worboys, 1995). A regular lattice of estimated values spaced 100 m apart was then extracted from the TIN and subsequently processed to identify lines of equal travel time (known technically as isochrones, see Brainard *et al.*, 1997).

13.3.2 Measuring Accessibility by Bus

Estimating bus journey times was not regarded as feasible or appropriate in this study and instead an approach was developed which sought to determine whether residents could walk to a bus route that would take them close to the nearest facility of a particular type. The procedure implemented within the Arc/Info GIS involved taking each surgery location in turn and first defining an 800 m radius buffer zone around it. A radius of 800 m (approximately half a mile) was used to represent a distance that the great majority of the population would find acceptable to walk. The GIS then selected all bus routes with a particular service frequency (*e.g.* at least four return journeys per day) that passed through the buffer zone. Next, a second buffer zone extending 800 m each side of the relevant routes was determined and the GP patient register postcodes within this corridor were identified. Residents with these particular postcodes were subsequently classified as populations, which had the relevant level of bus access to the surgery being examined. The procedure was repeated for all surgeries in both years using the two categories of route where there was at least one daytime return service every weekday.

There are clearly some limitations in the method described above. The selection of a maximum walking distance of 800 m was inevitably arbitrary, but reflected advice from local bus operators and transport planners. Location of bus stops were essentially ignored, but most services would stop in towns and villages and it is common practice to flag down a bus from the roadside in many rural areas. It would also have been possible to extend the analysis to include less frequent services, but it was decided that it would be more appropriate to use the parish classifications to represent a lower level of bus availability (*e.g.* this allows for longer walking distances). All these considerations mean that the results obtained need to be interpreted with care, but several of the simplifications counterbalance each other and the procedure is certainly sufficiently robust to identify general contrasts in accessibility by bus.

13.4 RESULTS

13.4.1 Travel Time to Nearest Surgery

In both Autumn 1997 and Autumn 2000 the average population-weighted car travel time from the patient register postcodes to the nearest GP surgery was approximately five and a half minutes. Maximum values in both years were just over 21 minutes. Figure 13.3 shows the pattern of isochrones for Autumn 2000, with shorter travel times occurring around the towns or along main roads, and longer journeys in a number of more rural areas. Table 13.2 confirms the skewed distribution of physical accessibility, around 58% of residents having a travel time under five minutes and 17% a journey longer than 10 minutes.

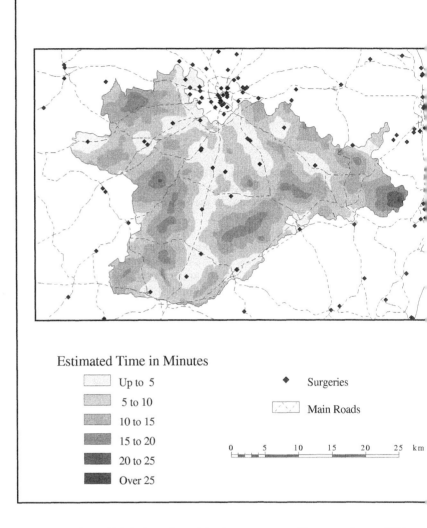

Figure 13.3 Estimated travel time by car to nearest GP surgery, Autumn 2000.

Table 13.2 Population by car travel time to nearest surgery.

	Car Travel Time Category			
Time Period	Under 5 Minutes	5 to 9.99 Minutes	At least 10 Minutes	Tot
Autumn 1997	62,277	27,522	18,515	108,
Autumn 2000	65,771	28,706	18,903	113,

13.4.2 Accessibility by Bus

Figure 13.4 shows the classifications of parishes by majority bus service in 1997 and 2000. As already acknowledged, the classifications conceal variations in provision within some parishes, but the maps certainly imply an improved situation by Autumn 2000. This is particularly apparent in terms of the reduction in parishes with the poorest levels of bus service provision.

Results from the more detailed assessment of accessibility to GP surgeries using the two most frequent categories of service are listed in Table 13.3. These figures indicate an increase in the number of South Norfolk residents with the most frequent category of bus service between 1997 and 2000, although the overall proportion of the district population covered by daytime return services every weekday remained stable at around 86 %. This implies an improved frequency of service, though it actually reflects a situation where improvements in many parishes were partly counterbalanced by declines in provision for a few larger villages. This point is illustrated in Figure 13.5 which shows the percentage of residents in each parish within 800 m walking distance of a weekday daytime return bus service for both 1997 and 2000. The trend is not especially pronounced, but a wider coverage of parishes by Autumn 2000 is nevertheless apparent.

Table 13.3 Populations with daytime return bus services to GP surgeries, 1997 and 2000.

Frequency of Bus Service each Weekday	Population		% of District Total	
	1997	2000	1997	2000
>= 4 Daytime Return Services	84,529	91,907	78.1	81.1
1-3 Daytime Return Services	9,122	5,483	8.4	4.8
Total	93,651	97,390	86.5	85.9

13.4.3 Availability of Community Transport

The possible impact of the CarLink driver distribution was examined by calculating the percentage of population in each parish who lived within three miles of the nearest volunteer driver. A three mile limit was used because the CarLink pricing structure involves a fixed minimum fee when the driver travels up to six miles, and a per mile rate beyond that distance threshold. It would therefore be more expensive for anyone living more than three miles from a CarLink driver to use the service.

Figure 13.6 displays the resulting percentage values. Most parishes have a high level of coverage (*i.e.* all residents are within three miles of a driver), but there are obvious gaps in some southern and eastern parts of South Norfolk. A 75% threshold was used to classify parishes as having reasonable proximity to a CarLink driver and these data were then combined with details of the catchments for other services to produce the overall categorisation of community transport provision shown in Figure 13.7. The results of this assessment are generally reassuring since they indicate that in Autumn 2000 there were no parishes without some form of community transport provision and many areas had at least two service providers. Table 13.4 lists the numbers of parishes and the size of the resident population for the different categories in Figure 13.7.

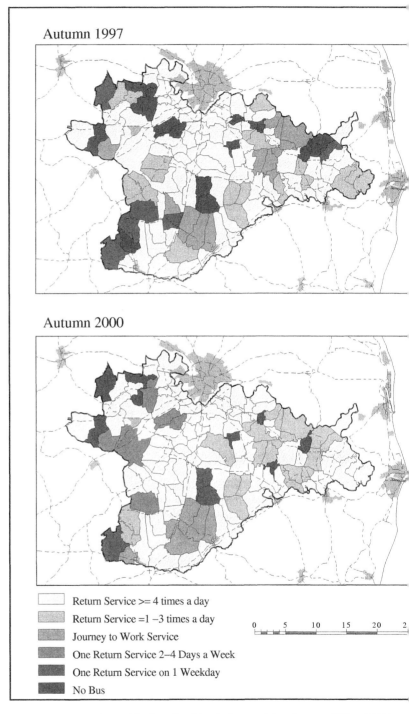

Figure 13.4 Classification of parishes by bus provision, Autumn 1997 and 2000.

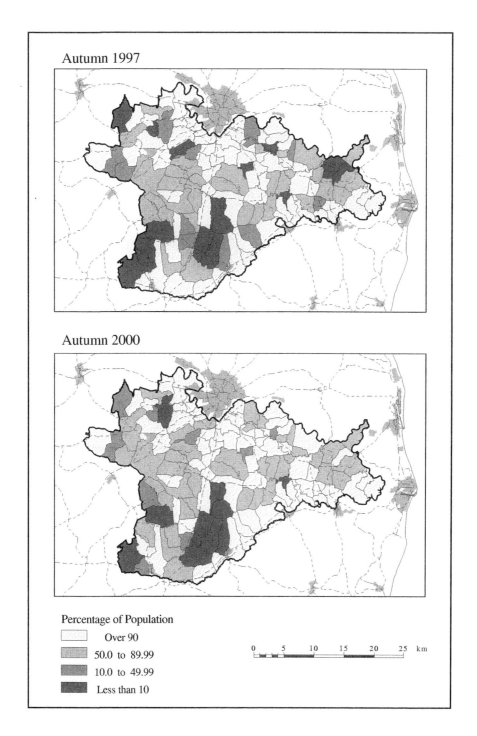

Figure 13.5 Percentage of residents within walking distance of a daytime return bus service.

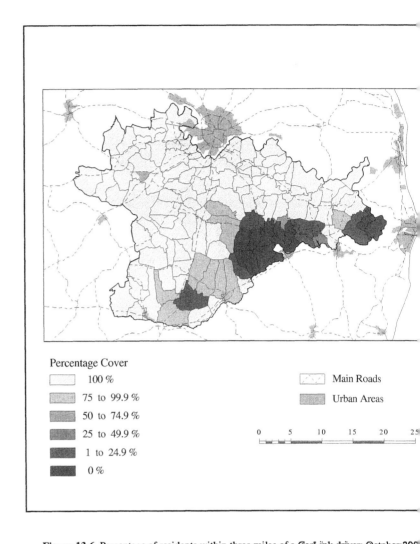

Figure 13.6 Percentage of residents within three miles of a CarLink driver, October 200

Although it is difficult to precisely quantify the change in community
services since Autumn 1997, there seems little doubt that the situation has impro
schemes have been established in a number of small parish groupings and the in
of the Wymondham Flexibus has provided additional services in a significant
rural parishes. It is worth noting, however, that organisations such as CarLink a
Community Transport are very dependent on the availability of volunteer driv
one parish was classed as having a 'Limited Carlink Service' (because fewer th
residents were within three miles of a driver), but a number of others could eas
this level if a few key volunteers were no longer available.

Table 13.4 Classification of parishes by community transport provision in Autumn 2000.

Community Transport Provision	Parishes	Population in Autumn 2000	% of District Total
Limited CarLink Service	1	1,400	1.2
CarLink Service Only	51	39,061	34.5
Bungay Dial-a-Ride Only	17	7,273	6.4
CarLink & Community Car Scheme	7	6,236	5.5
CarLink & Another Dial-a-Ride	40	57,797	51.0
CarLink & Two Additional Services	3	1,613	1.4

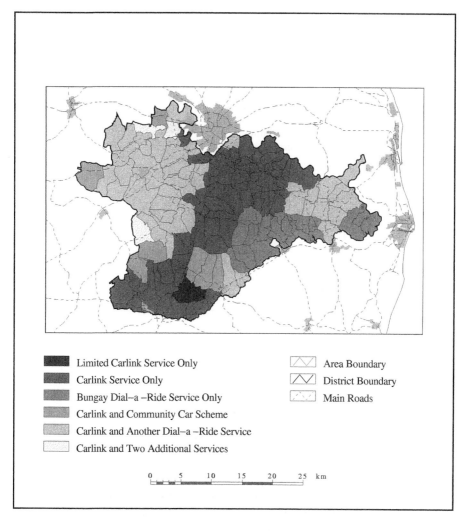

Figure 13.7 Classification of parishes by community transport provision, Autumn 2000.

13.4.4 Overall Accessibility by Private and Public Transport

In order to examine the overall improvement in accessibility, the postcoded dat
year were crosstabulated in terms of car travel time and the availability of bu
Table 13.5 presents the result for Autumn 1997, the district population being s
into three car travel time bands and five bus service categories. The first two o'
represent populations living close to the most frequent buses, while the third c
residents more than 800 m from a route, but within a parish classed as having
return service each weekday. Category four contains populations over 800 m fr
route but in a parish with a less frequent bus service (*e.g.* on only one or t
week), while the final group are residents of parishes classed as having no bu
The progression down the rows of the table therefore reflects a decreasing acc
services.

Table 13.5 Population by car and bus accessibility to nearest GP surgery, Autumn 199

Bus Service Provision	Car Travel Time Category			
	Under 5 Minutes	5 to 9.99 Minutes	At least 10 Minutes	T
Within 800 m of Route with >= 4 Return Daytime Services every Weekday	58,430	20,544	5,555	8₄
Within 800 m of Route with 1 to 3 Return Daytime Services every Weekday	1,880	1,186	6,056	⁹
In Parish with Return Daytime Services every Weekday	680	2,791	1,835	⁵
In Parish with Less Frequent Bus Service	1,134	1,829	3,626	⁶
In Parish with No Bus Service	153	1,172	1,443	₂
Total	62,277	27,522	18,515	10⁸

One interesting feature of Table 13.5 is that all 15 cells of the crosstabu
occupied. This emphasises the point that car travel time and availability of bu
can be rather different dimensions of physical accessibility and also high
variations in accessibility within South Norfolk. Table 13.6 presents
crosstabulation based on the Autumn 2000 results and comparison with the
indicates a particular shift in population away from the poorer accessibility c
One illustration is provided by examining the totals within the over 10 minutes
time columns. Another approach is to regard the four cells in the bottom-right
the tabulation (*i.e.* those with car travel times of at least five minutes, no nearby
routes, and in parishes without a daytime return service each weekday) as popu
whom physical accessibility could begin to become more difficult. In Autumn 1
were 8,070 South Norfolk residents in these categories and by Autumn 2000 the
declined to 6,945. Of the latter, 54% were in parishes where at least two c

transport schemes were operating and there were only 93 people in areas more than 10 minutes car travel time from a GP surgery, without a bus service, and with only one community transport provider.

Table 13.6 Population by car and bus accessibility to nearest GP surgery, Autumn 2000.

Bus Service Provision	Car Travel Time Category			
	Under 5 Minutes	5 to 9.99 Minutes	At least 10 Minutes	Total
Within 800 m of Route with >= 4 Return Daytime Services every Weekday	61,574	21,021	9,312	91,907
Within 800 m of Route with 1 to 3 Return Daytime Services every Weekday	801	1,082	3,600	5,483
In Parish with Return Daytime Services every Weekday	665	2,833	2,816	6,314
In Parish with Less Frequent Bus Service	2,674	3,450	2,255	8,379
In Parish with No Bus Service	57	320	920	1,297
Total	65,771	28,706	18,903	113,380

The structure of the patient register data also makes it possible to examine the distribution of population with potential accessibility difficulties by GP practice. Table 13.7 lists the percentage of registered patients in the four poorer accessibility categories discussed above for a sample of South Norfolk practices in Autumn 1997 and 2000. The contrasts between practices and over time are evident, demonstrating the usefulness of such information for primary health care planning and management.

Table 13.7 Percentage of population with potential accessibility difficulties for selected GP practices.

GP Practice	Percentage of Registered Patients in Poorer Accessibility Categories	
	1997	2000
A	14.0	1.0
B	0.2	0.2
C	3.6	2.4
D	10.0	4.9
E	6.5	5.8
F	25.7	23.6
G	7.0	8.0

13.5 CONCLUSIONS

The results of the research described in this chapter indicate a broad impro▮ levels of accessibility to GP surgeries between Autumn 1997 and Autumn 2C was relatively little change in travel times by car, but more residents had higher bus routes within walking distance and there was enhanced bus provision in a the remoter rural parishes. By Autumn 2000 there were community transpor covering all South Norfolk parishes and over half the district population had a▮ service providers available.

These findings suggest that the public sector investments in rural trans▮ early 1998 have had a definite beneficial impact on accessibility. There are substantial contrasts in accessibility to GP surgeries within the district, but it ap the populations for whom travel to facilities remains difficult are now sn▮ dispersed across many parts of the district. The best means of helping such r▮ likely to be through further expansion of community transport, perhaps th▮ creation of new schemes or the recruitment of additional volunteer drivers in ar availability is currently limited. There may also be a case for some inve information dissemination and co-ordination, particularly so that residents knc request such transport services.

Although the period since early 1998 has seen the introduction of many n transport services, it is fair to note that actual usage has sometimes been disa▮ This, in turn, has lead to adjustments in timetables and routes where, if anyt probably stability in service provision that would help to encourage greater us increases in fuel costs have also created problems, with several bus ope organising routes to provide a greater frequency of service between towns, detours to villages away from main roads (Clement, 2000; Hill, 2000). This n some of the results presented in this chapter are already out-of-date. More posi▮ Rural White Paper published in November 2000 proposed further funding transport (see `http://www.wildlife-countryside.detr.gov.uk/ruralwp/ir` The new Parish Transport Grants scheme may offer particular opportunities fc such as South Norfolk as it explicitly mentions support for community transp existing bus services to divert through a village.

The results presented in this chapter have focused on one local authorit recent experience of South Norfolk is likely to be reflected in many other low areas of the UK. The research certainly demonstrates the analytical power pr▮ combining GIS techniques with data sources such as patient registers and di networks, and the methodology presented in this study could certainly be applie locations. It must be acknowledged that the work required to produce digital information was time consuming, and it is also important to recognise that confidentiality issues associated with the use of patient register data. Studies su nevertheless demonstrate the considerable research resource that patient regis▮ provide to the NHS, and it is to be hoped that developments such as the establi▮ Regional Public Health Observatories (Department of Health, 1999) will hel▮ anonymised extracts or summaries more widely available for planning and ma▮ purposes.

13.6 ACKNOWLEDGEMENTS

The research described in this chapter was undertaken during parts of projects South Norfolk Council, Norfolk Rural Community Council and the NHS

Eastern Region. We would like to thank Alexandra Bone for her assistance in compiling the information on bus services, and the following for their help with data or guidance: Jen Wingate and Jeananne Waugh, South Norfolk Council; Lyn Reynolds, South Norfolk Primary Care Group; Julian Bester, East Norfolk Health Authority; and Andrew Coleman, Liz Joyce and Wendy Pontin, Norfolk County Council. For census products and digitised boundaries, we acknowledge the source: The 1991 Census, Crown Copyright, ESRC Purchase.

13.7 REFERENCES

Bateman, I.J., Lovett, A.A. and Brainard, J.S., 1999, Developing a methodology for benefit transfers using geographical information systems: modelling demand for woodland recreation. *Regional Studies*, **33**, pp. 191-205.

Brainard, J.S., Lovett, A.A. and Bateman, I.J., 1997, Using isochrone surfaces in travel-cost models. *Journal of Transport Geography*, **5**, pp. 117-126.

Clement, B., 2000, Bus firms blame high cost of fuel for scrapping of routes. *The Independent*, 28th November 2000.

Countryside Agency, 2001, *State of the Countryside 2001*, (Cheltenham: Countryside Agency).

Department of Environment, Transport and the Regions, 2000, *Our Countryside: The Future – A Fair Deal for Rural England*, (London: The Stationary Office).

Department of Health, 1999, *Saving Lives: Our Healthier Nation*, (London: The Stationary Office).

Department of Transport, 1993, *Vehicle speeds in Great Britain 1992*, Statistical Bulletin **93**(30), (London: Department of Transport).

Gatrell, A.C., Dunn, C.E. and Boyle, P.J., 1991, The relative utility of the Central Postcode Directory and Pinpoint Address Code in applications of geographical information systems. *Environment and Planning A*, **23**, pp. 1447-1458.

Haynes, R.M., 1986, *The Geography of Health Services in Britain*, (London: Croom Helm).

Haynes, R.M., Lovett, A.A., Bentham, C.G., Brainard, J.S. and Gale, S.H., 1995, Comparison of ward population estimates from FHSA patient registers with the 1991 Census. *Environment and Planning A*, **27**, pp. 1849-1858.

Higgs, G. and White, S.D., 1997, Changes in service provision in rural areas. Part 1: The use of GIS in analysing accessibility to services in rural deprivation research. *Journal of Rural Studies*, **13**, pp. 441-450.

Higgs, G. and White, S.D., 2000, Alternatives to census-based indicators of social disadvantage in rural areas. *Progress in Planning*, **53**, pp. 1-81.

Hill, P., 2000. Give and take on the buses. *Eastern Daily Press*, 20th November 2000.

Jones, A.P. and Bentham, G., 1997, Health service accessibility and deaths from asthma in 401 local authority districts in England and Wales, 1988-92. *Thorax*, **52**, pp. 218-222.

Joseph, A.E. and Bantock, P.R., 1982, Measuring potential physical accessibility to general practitioners in rural areas: a method and case study. *Social Science and Medicine*, **16**, pp. 85-90.

Joseph, A.E. and Phillips, D.R., 1984, *Accessibility & Utilization: Geographical Perspectives on Health Care Delivery*, (New York: Harper & Row).

Lovett, A.A., Haynes, R.M., Bentham, C.G., Gale, S., Brainard, J.S. and Sünnenberg, G., 1998, Improving health needs assessment using patient register information in a GIS.

In *GIS and Health*, edited by Gatrell, A.C. and Loytonen, M. (London: Francis), pp. 191-203.

Lovett, A.A., Haynes, R.M., Sünnenberg, G. and Gale, S., 2000 *Accessibility c Health Care Services in East Anglia*, Research Report No. 9, (Norwich: Health Policy and Practice, University of East Anglia)

Martin, D., 1992, Postcodes and the 1991 Census of Population: issues, pro prospects. *Transations Institute of British Geographers*, **14**, pp. 90-97.

Moseley, M.J., 1977, *Accessibility: The Rural Challenge*, (London: Methuen).

Naude, A., de Jong, T. and van Teeffelen, P., 1999, Measuring accessibility tools: A case study of the wild coast of South Africa. *Transactions in GIS*, **3** 395.

Norfolk County Council, 2000, *Mid-1999 Small Area Estimates for Norfolk Urban Wards, County Electoral Divisions and Built-Up Areas*, Dei Information Note 5/00, (Norwich: Planning and Transportation Departmen County Council).

O'Sullivan, D., Morrison, D. and Shearer, J., 2000, Using desktop GI! investigation of accessibility by public transport: an isochrone approach. *Int Journal of Geographical Information Science*, **14**, pp. 85-104.

Ricketts, T.C. and Savitz, L.A., 1994, Access to health services. In *Geographi for Health Services Research*, edited by Ricketts, T.C. *et al.*, (Langham, : University Press of America, Inc.), pp. 91-112

Simpson, S., editor, 1998, *Making Local Population Statistics: A Guide for Pra* (Wokingham: Local Authorities Research and Intelligence Association).

Wallace, M. and Denham, C., 1996, T*he ONS Classification of Local ar Authorities of Great Britain*, Studies on Medical and Population Subject (London: HMSO).

Worboys, M.F., 1995, *GIS: A Computing Perspective*, (London: Taylor & Fran(

14

Measuring accessibility for remote rural populations

Mandy Kelly, Robin Flowerdew, Brian Francis and Juliet Harman

14.1 INTRODUCTION

In the UK services such as Police, Fire, Education, Social Services and Health are supplied by government to the population. Funds are generated locally (through means such as Council Tax) and are topped up by Central Government to provide a standard level of service. The way in which funds are allocated varies across the UK but in this paper we are focusing on England, where funds for services (except for health) are allocated by the Standard Spending Assessment (SSA) formula. The Standard Spending Assessment system was devised to determine how allocations should be modified from a strictly per capita basis. The system is based mainly on a regression analysis of costs incurred by authorities on a set of explanatory variables that might seem relevant to costs.

This paper demonstrates the calculation of a set of new measures of population remoteness, using GIS techniques to measure travel distance and time on the road network, and examines the results of the incorporation of these measures into the existing SSA formula.

14.2 BACKGROUND

The premise of this work is the argument that the cost of service provision is strongly affected by population distribution (Shropshire County Council (1996), Northumberland County Council (1997), Hale and Capaldi, (1997)) and it is the development of earlier work by Flowerdew and Gill (1998), which used straight line distance to create similar measures. The assumption is that if the service involves travel, either by clients to service delivery points or by service workers to the people, the distances travelled are likely to be longer in a rural area than an urban one, as are travel times, and the travel costs will be higher.

The most common approach to measuring population remoteness is through calculating a sparsity measure. At present sparsity measures are included in SSA calculations for some services but not for others, and are included in different ways for different services (see DoE, 1997). The measure of sparsity of population used in SSA calculations seeks to take account of both the overall sparsity in an authority and areas of

extreme sparsity within authorities that are not generally so sparsely popula█
Education SSA, it is based on a weighted sum of 'ordinary sparsity' and 'supe█
Ordinary sparsity is defined as the proportion of the authority's population th█
wards where the density of population is between 0.5 and 4 people per hecta█
sparsity is defined as the proportion of the authority's population that lives█
where the density of population is 0.5 or fewer people per hectare. Super-sparsi█
double the weight of ordinary sparsity. For EPCS services, the definition is█
except that calculations are based on enumeration district (ED) rather than
(DoE 1997).

It is argued that sparsity, as defined above, is not a good measure of the
costs incurred in delivering services to a rural area (Flowerdew and Gill, 1998█
does not take into account the geographical distribution of the population, in
the relative location of the population being served and the centres from which █
is provided. Thus a sparsely populated ward or ED close to a service deliv█
would not be nearly as expensive to serve as one with a similar degree of spars█
far away from the centre. A better measure would be based on the distance or █
between the service delivery centre and the population.

Until recently, such a measure would have been time-consuming and █
construct. However, with the rapid advance of geographical information syst█
technology and suitable data sets, we hoped to supply a range of measures of th█
to try them out in the SSA regressions. Similar approaches have been taker█
recent studies attempting to assess accessibility in different contexts, such a█
Lovett *et al.* (2000) on accessibility of primary health care services in East A█
Higgs and White (1997) on accessibility to services in Wales.

This paper will provide an overview of the GIS techniques used to derive
alternative measures of population remoteness for local government areas using
travel times and distances from each ED to assumed service delivery points,█
different assumptions about where the services are located and on the local auth█
which the measure was to be computed. Once these measures had been const█
second objective was to examine the impact of using them (rather than the co█
sparsity measures) in the calculation of Standard Spending Assessment (S█
methodology of this objective is not outlined in detail here, although some o█
results are discussed.

14.3 METHODOLOGY

The construction of measures of remoteness required the use of a GIS to cal█
lengths of the shortest paths along the road network between each ED centro█
nearest service delivery point. This was done using Bartholomew's 1991 1:250█
digital map database of the road network obtained under the CHEST agree█
populations and grid references for ED centroids were obtained from the MIM█
at Manchester University, using data purchased by ESRC and JISC on beh█
academic community. Accessibility measures were computed as if all trips we█
or from the population weighted centroid of the ED, or the nearest point or█
network.

A considerable amount of time was necessary for constructing coverages █
system. One unanticipated problem was the necessity to specify locations

limited-access roads like motorways could be joined – otherwise the algorithm would allow traffic to join or leave wherever the limited-access road was bridged by another road. Routes were not constrained to be completely within the local authority for which the measures were being calculated. Thus it was possible for a route from an ED to a service delivery point within the same authority to pass through a different authority if distances or times were reduced by so doing.

Calculation of travel times required estimates to be made of the speeds of travel on roads of particular quality. Standard estimates of travel times, such as the national off-peak estimates produced by the Department of Transport (1993), seemed inappropriate in the context of travel much of which would be done at or near peak hours. Instead, we used average speeds computed for four types of roads based on field testing in North West England, conducted in order to determine accessibility for private sector service provision. These are 46.55 mph on motorways, 28.60 mph on A roads, 22.20 mph on B roads, and 15.13 mph on other roads. These speeds are intended to be averages reflecting different degrees of traffic density, different driving styles, and variable effects of weather, congestion, joining and leaving delays, speed limitation and traffic lights. the speeds themselves are less important than the ratios between the speeds obtained for different types of road.

Islands were excluded from the analysis. In many cases they are not dealt with appropriately in GIS coverages of the census data. A small offshore island may be part of a mainland ED with the same ED code assigned to it, and may incorrectly be assigned the total population for that ED. The census data do not provide information on how the population is split geographically within the ED, and it is not possible to tell from the data whether an island is inhabited or not. In addition, if an island is not connected to the mainland as part of the road network, it would be necessary to assign some arbitrary measures of travel distance and time.

Initial experiments had investigated the use of ArcInfo, as functions exist to calculate the distance between nodes on the network (ArcInfo command NODEDISTANCE). Each ED and service centre could be allocated to a node (using ArcInfo command NEAR). However, it was found that the robustness of the road network was seriously limiting. Any change to the road network would result in a renumbering of the nodes and all the preparation stages had to be repeated. There were also problems with nodes that had an ID of zero, due to truncation or clipping of the network, and these were indistinguishable from each other. The use of Bartholomew's data may have been an influencing factor but even with the more accurate Ordnance Survey road data (such as OSCAR) the work may not have been problem-free.

Due to the problems encountered in ArcInfo, an extension available for ArcView (Shortest Network Paths v1.2 developed by Neudecker (1999)) was used to calculate the travel distance and time. This associated points with nodes on the network, identified all the possible paths along the network and then found the minimum cost (either distance or time). Since this software existed and was generally suitable and in ArcInfo further programming would be necessary to facilitate automation, the analysis was continued in ArcView. The extension was further adapted using Avenue programming language to aggregate the measures, weighted by ED population, to produce a single measure for each local authority. This measure could be interpreted as the total travel distance (or time) needed for each resident to reach the nearest service delivery point. If desired, this could be divided by population to get a per capita measure of remoteness that would be comparable between local authorities of different size. Similar computations ⌐ using time rather than distance as the basis of calculation.

Because different levels of local government deliver different servic⌐ appropriate to reproduce the measures for different types of authority. Al⌐ included in SSA are delivered by Metropolitan Districts (including London and Unitary Authorities. Some services, including education and Personal Soci⌐ (PSS), are delivered by the shire counties while others, including Envi⌐ Protective and Cultural Services (EPCS), are delivered by districts within counties. Different databases were therefore necessary for consideration o⌐ services. To examine services at an upper tier level (Metropolitan District⌐ Authorities and shire counties, see Figure 14.1) we looked at the Other Perso⌐ Services component of the PSS Standard Spending Assessment. To investigate of new measures of remoteness for authorities at the 'lower tier' level (M⌐ Districts, Unitary Authorities and districts within shire counties, see Figure 14⌐ were also examined.

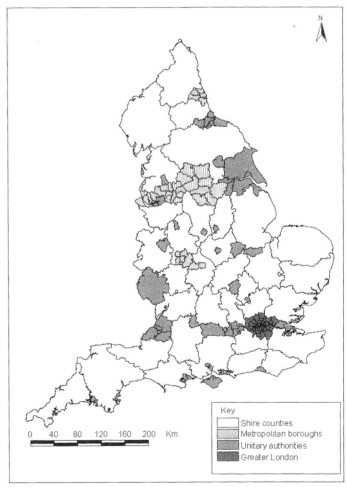

Figure 14.1 New upper tier local government geography of England 1998.

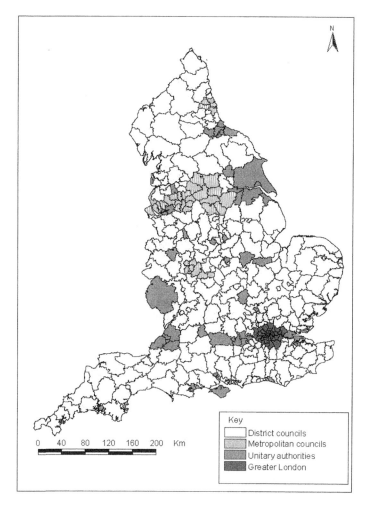

Figure 14.2 New lower tier local government geography of England 1998.

In addition, the regressions used in the development of SSA were based on calculations performed using the local authorities in existence before the introduction of unitary authorities in the late 1990s. Accordingly the measures were also calculated for the 'old' local authorities, in which unitary authorities were included within the county in which they were located. In practice these calculations were actually easier than those for the 'new authorities' as some boundary changes could not be constructed easily directly from the previous system. Population data for these new boundaries were supplied by look-up tables by Wilson and Rees (1999).

The pattern of service provision is likely to differ substantially between different service sectors. In particular, the number of service delivery points may differ, and be either clustered or dispersed and single or multiple.

A series of meetings with local authorities was scheduled, intended to get their views about the role of geographical distance in the organisation of services and their costs. Respondents were chosen to give a representative selection of different types of

local authority. The selection was also influenced by the responsiveness
authorities invited to take part. Interviews were held with representatives of I
County Council, Barrow District Council, Norfolk County Council a
Lincolnshire Unitary Authority. They were asked a series of open-ended quest
the organisation of services, including issues concerned with the size of settl
which a particular service might be available, and with the role of distance
provision.

It is clear from these examples that remoteness and accessibility are im
these authorities, not just in terms of the distances that must be travelled for
meet service suppliers, but in a range of more subtle ways. The interv
strengthened our view that distance and travel time measures are more rele
sparsity. Unfortunately the situation is complex and the size of the effects is c
measure.

To model some of these possibilities we used the following:

1.   the (population-weighted) centroid of the local authority only. The
      centroid was used as it is the simplest model (although somewhat unrea
      enabled the development and assessment of the efficacy of the methodolog

2.   the nearest district centroid. These measures were equivalent for Me
      Districts and Unitary Authorities (which could be regarded as single dis
      differed substantially for counties, where service delivery from a single po
      most likely to lead to very long travel distances.

3.   distance to the nearest settlement of a given size (using ONS definition
      rural areas).

14.4 RESULTS

There are three main areas of results. Each analysis creates a set of aggreg
distances and times for each local authority (see Figures 14.4 to 14.6). The
compared with the sparsity measures currently used in SSA calculations (Fig
The redistribution of resources that would be implied by substituting the r
measure for sparsity can also be computed, to identify the 'winners' and 'losers'

The number of sets of measures produced was not as extensive as initia
There were several contributory factors during the data preparation and ana
machine ran to capacity whilst calculating the measures. Individual runs were
consuming to the extent that there were major delays in completing this p
analysis, finally forcing us to reduce the number of measures for which r
indices could be calculated. In particular, despite our intention to produce me
distances and times to settlements of a range of threshold sizes, time only perm
calculations for a threshold size of 100,000.

Complications are introduced by the availability of data for some set of
units but not others. It is more useful to display the comparisons between the
sparsity measure and our measures of travel distance and time for exis
government units. However recalculation of the regression equations used for de
SSA formulae should be done using the old units (both for comparability, an
some of the data are only available for these units). Accordingly, we compute

and time measures for accessibility to local authority centroids, district centroids, and settlements of 100,000 population separately for the four sets of local authorities – the old upper tier, the new upper tier, and old lower tier and new lower tier. This was a considerable increase in the workload from what had originally been envisaged.

One problem that arose in incorporating our new measures into SSA methodology is that detailed information on recent changes in the methodology was not available until late in the project. Although we are indebted to DETR staff for supplying the information necessary to replicate their calculations and hence to examine the effects of our measures, problems were encountered in understanding exactly what had been done and in reproducing the DETR's results to sufficient accuracy.

It should be noted that most of the data in the SSA calculations are taken from the 1991 census, and hence are nearly ten years old. There is also a minor difference in that the population data for the 'old' areas are based on the 1991 census, while the data for the 'new' areas are based on a 1991 mid-year estimate. This leads to small differences in population totals even for areas whose boundaries have been unchanged in the reorganisation, and also to small changes in the locations of their population-weighted centroids.

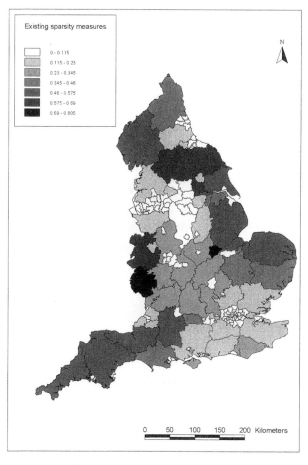

Figure 14.3 Existing sparsity measure.

14.4.1 Measures Computed

The maps (Figures 14.3 to 14.6) show the values of the various measures of r
used in the analysis. Figure 14.3 shows the existing sparsity measure
calculations at the ED level, mapped for the upper tier of authorities as
constituted. Generally the pattern is a reasonable reflection of remoteness, thou
be surprising that areas like Rutland and Shropshire appear to be more sparsely
than Northumberland, Cumbria, East Anglia and the South West.

Following is a series of maps showing the measures we have computed for
tier. There are three pairs of maps, representing an average distance measu
average travel time measure under three sets of assumptions. First, distance
time to the authority centroid are calculated (Figure 14.3). Generally there is a f
relationship between the distance and travel time results (with some
differences). In comparison to the sparsity map, these maps show a strong re
with the area of the county. Kent, for example, has a fairly high value bec
assumed that a fairly dense population has to travel considerable distances to
point in a large county. It may be noted however that Kent is less prominent in
time map, presumably because the motorway network makes the centroid more
than is the case for other large counties with less motorway mileage.

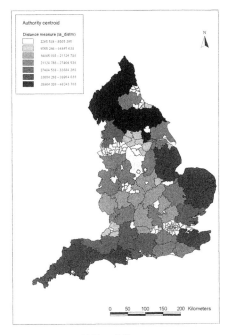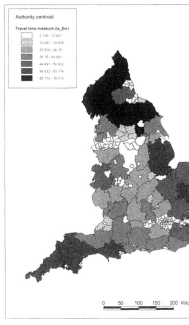

(a) Travel distance.   (b) Travel time.
Figure 14.4 Measures from EDs to the local authority centroid.

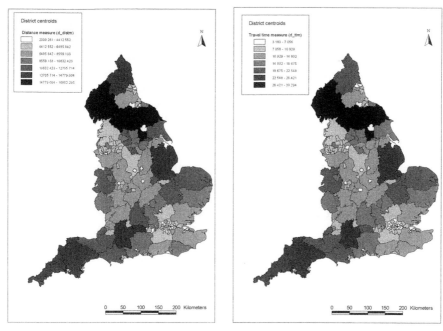

(a) Travel distance.   (b) Travel time.
Figure 14.5 Measures from EDs to the district centroid.

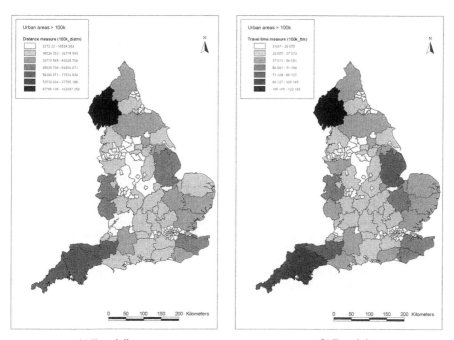

(a) Travel distance.   (b) Travel time.
Figure 14.6 Measures from EDs to urban areas over 100,000 population.

The second pair of maps (Figure 14.5) shows distances and travel tin nearest district centroid. These will be identical for metropolitan district boroughs and unitary authorities but will be shorter for the shire counties. sparsely populated unitary authorities, like Herefordshire and Rutland, prominent in the district maps while less sparsely populated but larger cor Dorset, Kent and Suffolk appear less prominent.

The third pair (Figure 14.6) shows distances and travel times to the nea area with a population of over 100,000, on the grounds that such places are like service facilities not present in smaller cities. The maps again show large and counties as having high values, with the most prominent being counties whi contain any large cities, such as Cornwall, Cumbria and Lincolnshire.

14.4.2 Comparision With Existing Sparsity

As one example of how our remoteness measures might impact calculations oi looked at the Other Personal Social Services component of the Personal Socia (PSS) Standard Spending Assessment ('Other' refers to aspects of PSS bes provided for children and older people). 'Upper tier' authorities (using the nor introduced above) provide these services. The client group is the popula between 18 and 64. The formula for the cost per client includes two indicators, by regression, that reflect social and health conditions in each authority by t account the difference in the cost of service provision.

The regression formula was successfully replicated to 3 significant figure remoteness measures were added, one at a time, as an extra explanatory variab regression formula. The effect of remoteness was negative in all regressions.

Table 14.1 Remoteness measures.

Remoteness Measure	Coefficient	Standard Error	t value	p
Sparsity (ED level)	-3.952	3.978	-0.99	0.
Distance to Local Authority Centroid	-0.0000899	0.0000432	-2.08	0
Travel Time to Local Authority Centroid	-0.04883	0.02647	-1.84	0
Distance to Urban Area of 100,000	-0.0000664	0.0000326	-2.04	0
Travel Time to Urban Area of 100,000	-0.04972	0.02592	-1.92	0

As it was not possible to calculate road distances to an urban centre of 1 more for the Isle of Wight, this authority is omitted from the last two regressic therefore use 106 observations rather than 107. Note that the relative si coefficients is dependent on the scale at which the measurements are conducted not reflect the magnitude of the effects, which is stronger for the new measures than for the traditional sparsity measure. The distance measures seem to perform slightly better than the travel time measures ($p = 0.04$), and there is little difference between the measure based on distance to the centroid and distance to the nearest large urban area. Only two of these five measures are statistically significant (distance to local authority centroid and distance of urban area of 100,000 plus $p = 0.04$).

14.4.3 Applying the New Measures to the SSA Formula

Because the effect of the remoteness measures in the regression analysis was negative when these figures were put in the SSA formula, the redistributive effects work against the interests of the remote rural areas. The largest gainers in percentage terms from introducing the distance to local authority centroid variable are Wokingham (13%), Rutland (10%) and Windsor and Maidenhead (8%); the biggest losers in percentage terms are Cumbria and Northumberland (-9%) and North Yorkshire (-8%). In absolute terms, Leicestershire and Bromley are the biggest gainers (over £500,000 per year) and Kent, Cumbria, North Yorkshire, Essex and Lancashire would all lose over £1 million per year. Using distance to the nearest urban area of over 100,000 gives the highest percentage increases to Wokingham (8%), Poole and South Gloucestershire (6%), with the largest percentage losses experienced by Cumbria (-21%), Cornwall (-13%) and Devon (-11%). The greatest absolute gains would accrue to Hampshire, Staffordshire, Surrey and Leicestershire, while the greatest absolute losses would occur in Kent, Cumbria, Devon, Cornwall, Lincolnshire and Somerset.

An overall conclusion from this analysis is that inclusion of a variable representing remoteness acts against the common-sense view that service delivery in the case of Other Personal Social Services would be more expensive in rural areas. The most likely explanation of this may be that the existing regression formula does not fully represent the costs relating to social conditions, which tend to be higher in the urban areas. The negative coefficient of the remoteness variable suggests that it may be acting as a surrogate for this urban effect. Only if the urban effect was adequately represented by variables more obviously reflecting it would the remoteness variable become positive and reflect the increased costs of providing services to a dispersed population in the way that was intended. It is interesting that our new measures reflect this effect better than the conventional sparsity calculation.

14.5 CONCLUSIONS

We feel that the methods used have been successful and that the measures produced represent a more sensitive approach to the problems of measuring remoteness than the sparsity index generally in use. Although we have not been able to complete all the analyses that we had intended to do in the project we have demonstrated that it is possible, though still time-consuming, to use GIS methods to construct a set of measures of accessibility for local authorities in England. However, we are not happy that the measures used are fully successful in capturing remoteness. In particular, they appear to give values that seem too high for large urban areas, where the assumption that services are provided from one centre may lead to excessive aggregate travel times and distances.

It may be appropriate to develop a method where the number of service assumed to be greater for urban areas with a larger population. Another alte modelling the pattern of service provision is to use the relative size of an within an authority, or to construct a database of actual service delivery pc gathering data from Local Authorities. An additional alternative is to apply a function to the travel times to allow for the difference in roads passing through rural areas.

The new measures have also failed to account for the discrepancy betwee expectations that service delivery costs would be greater for rural areas and results from the SSA regression analyses, which are inconsistent at best and i tend to show the reverse. Our analysis of Environmental, Protective and Cultura SSA indicates that the judgmental allocation of SSA based on sparsity is too compared to the results of including sparsity in the regression analysis. It is als that the sparsity measure is more significant in the SSA regression than our t and distance measures. In our analysis of Other Personal Social Services SSA, and distance are more significant than is sparsity, but the relationship is ne; service delivery costs appear to be lower for remote areas! This may be be measures do not deal adequately with large urban areas. However it seems n that this counter-intuitive relationship reflects a failure of other variables in the account adequately for socio-economic variables leading to costs being highe areas. A further possibility is that the relationship between costs and remotene non-linear. There is some evidence to suggest that costs are higher for the me and the least remote areas, and lower for intermediate areas.

14.6 FUTURE RESEARCH PRIORITIES

We would like to do more work on the measures, in particular on extending th for a greater range of settlement sizes and investigating different alternatives to the pattern of service delivery. We had hoped to do this during the project, b have sufficient time. There is also a problem with the assumptions ma methodology about service provision in large cities. This was recognised project and meetings with local authorities proved that to quantify the pattern provision for the whole country would be a difficult task. Accessibility app more of a problem according to our measures than seems likely in practice because we assume that accessibility is calculated from the district centroid, the district concerned has a large enough population to support several servic centres. We would like to develop methods for dealing with this problem, and more minor technical problems, before presenting our results as ready for policy arena.

Some of our results suggest that the relationship between remoteness a delivery costs may be non-linear. It appears that costs may be higher in remote but also in inner cities, perhaps as a result of congestion. We would like to e> possibility in more detail, both through further statistical analysis and throu discussions with local government service providers.

The nature of the system for central government support of local g activities is currently under review. It is generally accepted that additional costs issues of rurality and remoteness are problematic in the current system, so we a contribute towards a more satisfactory way of taking account of them. We would also be interested in reviewing how similar issues are treated in other countries (including the other component countries of the United Kingdom).

Similar considerations are also relevant to the provision of health services, where it is clearly important for everybody to be within reach of medical facilities. Rurality and remoteness are therefore again likely to be crucial in the funding formula for devolution of central funds to local areas. Current National Health Service funding takes this account in the funding of ambulance services but there is a strong case for it to be more prominent in other aspects of health service funding.

14.7 REFERENCES

DOE, 1997, *Standard Spending Assessments: Guide to Methodology 1997/98*. Local Government Finance Policy Directorate, (London: Department of the Environment).

Flowerdew, R. and Gill, M., 1998, Geographical measures of accessibility for modelling costs of service delivery. Paper presented at the *Geocomputation 98*, Bristol.

Hale, R. and Capaldi, A., 1997, *Local Authority Services in Rural England.* Rural Development Commission.

Higgs, G. and White, S. D., 1997, Changes in service provision in rural areas . Part 1: The use of GIS in analysing accessibility to services in rural deprivation research. *Journal of Rural Studies*, **13**(4), pp. 441-450.

Lovett, A., Haynes, R., Sunnenburg, G. and Gale, S., 2000, *Accessibility of Primary Health Care Services in East Anglia*. Research Report 9, School of Health Policy and Practice, (Norwich: University of East Anglia).

Neudecker, 1999, *Shortest Network Paths V1.2*. Developed from v1.1 by Kevin Remington, available from ArcScripts on `http://www.esri.com`.

Northumberland County Council, 1997, *Sparsity and the SSA*. Finance Department, Northumberland County Council.

Senior, M.L., 1994, The English Standard Spending Assessment system: an assessment of the methodology. *Environment and Planning C: Government and Policy*, **12**(1), pp. 23-52.

Shropshire County Council, 1996, *SSA and Sparsity: Executive Summary*.

Wilson, T. and Rees, P., 1999, Linking 1991 population statistics to the 1998 local government geography of Great Britain. *Population Trends*, **97**, pp. 37-45

15

Assessing the transport implications of housing and facility provision in Gloucestershire

Helena Titheridge

15.1 INTRODUCTION

With the rapid housing development expected to take place over the next decade, there is a real opportunity to place new housing in such a way that it supports existing or new services and facilities.

The hypothesis that allocating housing and other development based on the level of facility provision could result in reduced travel and fuel consumption was tested by applying a transport model to Gloucestershire. A number of development strategies based on balancing population and facility provision were modelled and the results compared with current travel patterns and those likely to result from housing allocations in the structure plan.

The greatest reductions were achieved through the strategy to expand as many towns as possible to above a threshold population size of 25,000. At this point a town was considered to be of sufficient size to support a number of higher order facilities as well as a wide variety of low order services and facilities. It is concluded that allocating housing to support service and facility provision could not only reduce travel but also significantly increase the level of access to key services and facilities for rural populations.

15.2 BACKGROUND

Planners in the United Kingdom are currently facing difficult decisions concerning the siting of new housing developments. UK Government projections predict that 3.8 million new dwellings will be needed between 1996 and 2021 (DETR, 1999). Pressure has been placed on local authorities to find suitable sites for this housing. Reducing traffic being just one of the considerations. Increasing access to employment, services and facilities reducing car dependency and social isolation being others. The later is particularly important in rural areas, where the level of service and facility provision if often poor. A recent survey by the Rural Development Council for the Countryside found that nationally

42% of parishes were without a permanent shop, 83% were without a genera
practitioner and 75% were without a daily bus service (CPRE: Council for the
of Rural England, 1998). This has a number of implications. Rural resider
travel longer distances than their urban counterparts, leading to a higher expe
fuel. CPRE (*op. cit.*) suggest that rural households spend an average of 10% m
on fuel than their urban counter parts, while ACRE (Action for Communitie
England, 1998) found that those living in rural communities travel 50% further
living in urban areas. For those with restricted or no access to a car the acce
services and employment opportunities can be extremely limited.

Government Planning Policy Guidance for Housing, PPG3 (DETR, 200
number of criteria for selecting suitable development sites:

- The availability of previously developed sites and empty or under-used bui
- The accessibility of local services;
- The capacity of existing infrastructure to absorb further development;
- The ability to build communities, to support new physical and social infr
  and to provide sufficient demand to sustain appropriate local services; and
- The physical constraints on development land.

Current emphasis in Government policy is to place new housing developr
on the edge of existing towns wherever possible. However, under certain circ
substantial sized development in villages are to be permitted along with new se
One of the criteria for village development is that "the additional housing w
local services, such as schools or shops, which could become unviable with
modest growth" (*op. cit.*, paragraph 70). Similar policies govern the developm
settlements, where new settlements must be of a size that can support a numb
services.

With the rapid housing development expected to take place over the ne:
there is a real opportunity to place new housing in such a way that it supports e
new services and facilities in line with PPG3. In order to achieve this Local A
need more information on the size of population and catchment areas needed
these services. Techniques are also needed to identify sites with development
based on this and the other criteria set out in PPG3.

This paper describes work undertaken as part of the URBASSS
Sustainability and Settlement Size) project, funded by the EPSRC Sustaina
programme. The aim of the research discussed in this paper was to tes
allocating housing and other development based on the level of facility provis
result in reduced travel and fuel consumption. This was done by applying the
(Estimation of Travel, Energy and Emissions Model) transport model to Glou
and comparing the travel patterns resulting from a number of development str
Gloucestershire based on balancing population and facility provision with cur
patterns in the region and with travel patterns likely to result from the housing a
given in the Gloucestershire Structure Plan (Gloucestershire County Council, 19

15.3 ESTEEM

ESTEEM was developed as part of EPSRC funded research carried out at UCI
1997 and 2000. The model assesses the sustainability of new developments i
personal travel demand, and associated energy consumption and emissions.
has been tested through application to Leicestershire and Kent (see Titherid
2000).

ESTEEM operates as an extension to ESRI's ArcView GIS package (Figure 15.1). ArcView was chosen because of its relatively low cost to Local Authorities, its network analysis capabilities and easy of customisation. An RTPI survey (RTPI, 1998) of Planning Departments, carried out in 1995, found that 7% of County/Region planning departments had access to ArcView. Over 30% had access to its sister product ArcInfo. A small survey conducted by the Bartlett School of Planning in 1997 (Titheridge *et al.*, 1998a) of County planning departments found a migration towards desktop pc and windows-based packages since the RTPI survey had been carried out. A more recent survey by the RTPI (RTPI, 2000) found that the number of Local Authorities using networked PC systems had increased from 50% in 1995 to 79% by 2000.

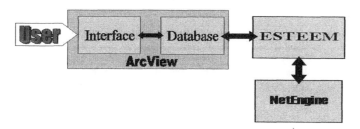

Figure 15.1 ESTEEM Structure.

An origin-constrained gravity model is used to simulate travel patterns by car, bus and rail for commuting, education, shopping, personal business and leisure trips. ESTEEM models approximately 80% of all motorised trips[1]. The modal split and the number of trips generated per person is determined by car ownership levels, age characteristics, employment levels, and the proximity to a bus route and to a rail station of each origin. The fuel consumption and emissions calculations take into account fleet characteristics, as well as estimates of cold start distances. See Titheridge and Rana (2000) for a fuller description of ESTEEM.

15.4 APPLICATION OF ESTEEM TO GLOUCESTERSHIRE

15.4.1 Data Input

Different population bases were used for calculating the trip rates for each type of journey purpose modelled, based on the demographic group most likely to make a trip for that purpose, see Table 15.1.

Table 15.1 Population bases for calculating trip production rates.

Purpose	Population
Work	Employees/Self-employed aged 16 or over
Education	Population aged 4 to 18
Leisure	Resident Population
Shopping	Resident Population

---

[1] Based on data from the National Travel Survey 1997/99 on journeys per person per year by main mode (DETR, 2000b).

Populations from the 1991 Census were used at enumeration district attributed to population centroids to give the data in the required point format, centroid acting as a 'trip origin'. Census data was used despite the fact that it years old because of its availability at such a fine scale and the completene sampling. Data on the fraction of the population living in households with a motor vehicle, was used to vary trip production rates for each mode, accord availability. Trip production rates were also varied according to public accessibility.

Destinations for each purpose were entered into the model as points. different level of aggregation was used for each purpose. These are listed belov 15.2, together with the source of the data.

Table 15.2 A summary of the data used in ESTEEM to represent trip destinations within Glouce

Purpose	Attraction Measure	Destination Aggregation	Source
Work	Total Employment	Ward	1991 Census
Education	Pupils on the School Roll	Individual School	DFEE
Leisure	Leisure Employment Index	Town Centres	Yellow Pages B Database
Shopping	Total Retail Employment	Town Centres	Yellow Pages B Database

The employment index used for the Leisure attraction measure was calcul: the following method. National travel survey data was used to derive the mean trips per year made per person to each leisure purpose type – to eat/c entertainment, to participate in sports, and for a day trip. Assuming that over 95 in each category are made to destinations within the study area and its buffer the trips per employee in the study area can be derived. Each business was al one of the four leisure purpose categories. The number of employees of that bus then multiplied by the appropriate trips per employee figure to give an attract for each business comparable between leisure types. The attraction figures aggregated by summation to town centres.

The 1991 census journey to work data for Gloucestershire shows that 9 work trips are to destinations within the county (OPCS, 1994). Assumin proportions for other journey purposes, then it is estimated that the model described above includes 52% of all trips made by the residents of Glouceste 73% of motorised trips, based on data from the National Travel Survey 97/9 2000b).

15.4.2 Calibration of the Model

Finding suitable data with which to calibrate the Gloucestershire model proved No suitable local travel surveys had been carried out. The 1991 Census of F included questions on travel but only on commuting trips to the normal place The National Travel Survey includes data on the county of residence for each re so trips by residents of Gloucestershire could be retrieved. However, the sample too small to produce statistically reliable information on a single county. solution was to calibrate the model using mean journey distances and travel tir on counties with similar characteristics to Gloucestershire.

Gloucestershire is predominantly a rural county. UWE (2000) found the following factors distinguished rural and urban areas: Population Density, Employment Density, Settlement Size and The percentage of the population in Agricultural Industries. For simplicity only two of these determinants were used in selecting 'like' counties – population density and the percentage of the population employed in agricultural industries.

Table 15.3 Mean Trip Lengths derived from National Travel Survey data 1988-95, for the calibration sample and for all counties and regions in Great Britain.

| | Mean Trip Lengths by Purpose (km) | | | |
	Commuting	Education	Shopping and Personal Business	Leisure and Social*
Calibration Sample				
Car	14.8	7.4	10.5	17.8
Bus	8.7	11. 2	7.3	42.1
All Counties				
Car	13.8	6.2	9.5	16.8
Bus	9.7	7.9	6.6	18.0

*Excludes trips to visit friends at their home, trips to a holiday base and just walk trips.

Using these two determinants, counties were selected if both parameters were within plus or minus a factor of two of the values for Gloucestershire. This gave a list of 24 counties including Gloucestershire. National Travel Survey (NTS) data for several years either side of 1991 was used to boost the number of trips included in the sample further. Trips and Journey Distance were then aggregated by mode and purpose to give mean trip distance for each trip type, *i.e.* commute journeys by car (Table 15.3).

The model was then run for each mode-purpose combination. In each case, the distance-decay exponent was adjusted until the modelled mean trip distance and the NTS mean trip distance converged. Because, the data used to calibrate the model, was not specifically for Gloucestershire and contained no further spatial disaggregation, it was not possible to assesses the statistical fit of the calibrated model at ward level, as done previously when calibrating ESTEEM (see Titheridge *et al.*, 1998).

Problems were encountered calibrating leisure trips by bus using the Gloucestershire equivalent sample. The NTS trip data includes all trips ending in the UK. The destination data included in the model however, is limited to locations within Gloucestershire. For most purposes, the percentage of external trips is small, so this is not too much of a problem. However, for rural areas like Gloucestershire with poor public transport provision, leisure trips by bus will tend to include a large proportion of coach excursions, with the associated long journey lengths. To get around this problem, leisure trips by bus were calibrated against NTS data for all counties rather than the smaller sample representing Gloucestershire equivalents.

15.5 CURRENT TRAVEL PATTERNS

After calibration the model input data was updated to represent the situation in Gloucestershire in 1998 (the most recent year for which population estimates were available). As more recent population estimates are not available at the enumeration district level or for all the variables used within the model, a number of assumptions about the changing level of car ownership and structure of the population had to be made (see Titheridge, 2000, for a full description of these assumptions). The model was then run for the entire county to establish current travel patterns.

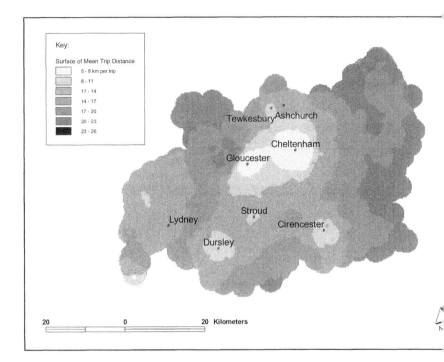

Key:

Surface of Mean Trip Distance

	5 - 8 km per trip
	8 - 11
	11 - 14
	14 - 17
	17 - 20
	20 - 23
	23 - 26

Figure 15.2 The Pattern of Mean Trip Lengths in Gloucestershire, 1998.

It was found that those living in the Severn Vale area tend to travel o
shorter distances than those living elsewhere in the county (Figure 15.2). T
surprising given that this region has the highest density of employment an
within the county, contains the towns of Gloucester and Cheltenham, and the n
services. Short mean journey distances are also found in the Forest of Dea
provision in this area is not huge but connections to the rest of the county ar
possibly the distances to the full range of services and facilities on offer are too
inhabitants make do with the limited range of facilities available locally.

The majority of travel in Gloucestershire is by car[2], accounting for 89% c
92% of the total distance travelled by residents of Gloucestershire and 97% c
consumed for travel (Table 15.4). Nationally, car travel accounts for 93% of ro
mileage, using the same base for journey purposes. The 6,000 km per annum
per capita for Gloucestershire also compares favourably with the 6900 km tra
person per year on average within the UK for the same range of journey n
purposes (DETR, 2000b). Although it should be noted that the figure for Glou
excludes journeys to destinations beyond the county boundaries. As already
the 1991 census showed that 9% of commuter trips by residents of Gloucester:
to destinations outside the county. Taking this into account, and the fact that
are likely to be longer than the average for the area, then travel in Gloucestersh
similar to the national average

---

[2] Car travel includes journeys by motorcycle, taxi and minicab.

Table 15.4 Annual modelled travel and energy for trips within Gloucestershire, 1998.

Purpose	Mode	Trips per Capita per annum	Annual Travel per Capita (km)	Annual Energy per Capita (kg)	Mean Trip Distance (km/trip)
WORK	CAR	88.88	1,331	76.1	14.98
WORK	BUS	8.28	73	1.1	8.78
EDUCATION	CAR	32.14	238	14.8	7.39
EDUCATION	BUS	12.59	141	2.3	11.17
RETAIL	CAR	212.74	2,242	85.9	10.54
RETAIL	BUS	22.29	163	2.6	7.32
LEISURE	CAR	92.81	1,671	46.7	18.00
LEISURE	BUS	6.97	127	2.0	18.19
TOTAL		*476.71*	*5,986*	*231.5*	*12.56*

15.6 MODELLING THE TRANSPORT IMPLICATIONS OF VARIOUS DEVELOPMENT OPTIONS

A variety of development options were tested for Gloucestershire, representing different methodologies for assigning housing, services and facilities. Option 1 was our interpretation of the Gloucestershire structure plan. Option 2 tackled under provision of services in Gloucestershire settlements. Option 3 added housing to those settlements that were over-provided in terms of services. Option 4 combined options 2 and 3, in an attempt to balance service provision and housing across the county. In option 5 housing was added to a small number of settlements, bringing them above population thresholds for medium to high order services. Lastly, a slightly different approach was taken for option 6 in which public transport provision was increased substantially.

A limited range of services and facilities were considered in each option. These services and facilities, namely primary schools, higher education colleges, theatres, leisure centres, pubs, supermarkets, convenience stores, clinics, banks and garages, were chosen to represent the range of activities a person undertakes and a range of high to lower order services. Williams (2000) established a range of threshold populations need to support each of these facilities using a variety of different techniques. A mid-range population threshold for each facility was used to determine the number of each of these key services that each settlement within the County could support. Settlements of less that a 1000 population were not considered. No employment was added in any option except for employment directly related to the services and facilities listed above. The characteristics of the population added through the new developments considered in each option, reflected the population currently resident in that area. Thus there is slight variation between the population totals for each option.

15.6.1 Option 1

This option tested the impact of the housing allocations given in the structure plan for Gloucestershire (Gloucestershire County Council, 1999) on transport. Local plans for each of the districts were used in conjunction with the structure plan to determine more

precise locations for the housing allocations. The structure plan policy is to
much of the housing allocation as possible within the Severn Vale area,
Gloucester, Cheltenham and Stroud. In addition, there is to be some devel
Tewkesbury/Ashchurch area and Cirencester. Within the Forest of Dean, hous
go to support the forest ring – a cluster of settlements including Lydney (Fores
1996). In addition, a number of villages were identified in the local plans as a
for small infill developments. It should be noted that at the time of analysis, m
local plans were under review or predated the structure plan, so the results of th
may not truly represent the travel implications of the structure plan.

Table 15.5 Annual modelled travel and energy for trips within Gloucestershire, Option 1 – based
plan housing allocations.

Purpose	Mode	Trips per Capita per annum	Annual Travel per Capita (km)	Annual Energy per Capita (kg)	Mean Dista (km/
WORK	CAR	92.45	1356	77.68	
WORK	BUS	8.75	25	0.40	
EDUCATION	CAR	33.18	239	14.92	
EDUCATION	BUS	13.00	39	0.63	
RETAIL	CAR	216.88	1814	72.36	
RETAIL	BUS	23.06	42	0.67	
LEISURE	CAR	94.63	1741	48.53	
LEISURE	BUS	7.18	79	1.27	
TOTAL		*489.14*	*5334*	*216.44*	

The structure plan strategy resulted in a small reduction in travel per cap
increased proportion of journeys made by bus (Table 15.5). This gives a rec
energy consumption per capita and per trip. These changes reflect the reductio
trip length and an increase in bus use across work, shopping and leisure
However, the structure plan strategy did result in an increased use of the car for
trips. This is possibly due to the nature of school bus provision; Glouc
education authority provide school buses in areas where the distance to th
appropriate school exceeds a set distance – this distance varies depending on tl
school. In areas where no school bus is provided, parents are more likely to es
children the entire length of the journey to school – often by car. Thus where th
is large to the nearest school, bus use prevails whilst those living closer to the sc
more heavily on the car.

The structure plan strategy had no impact on the spatial pattern of mean tri
As no attempt had been made to model the employment or transport polici
structure plan, the proportion of trips made by car as opposed to bus remained u
and the choice of destinations and accessibility of them was also unchanged.

15.6.2 Option 2

Option 2 aimed to test the implications for travel of boosting the number of ser
facilities in those settlements that our research showed were under provide

number of each facility within the urban area was lower than a settlement of that size could support. For each settlement that was found to have underprovided in one or more of the nine services studied, facilities were added to the shopping, leisure and education themes to bring the number of each facility type up to the expected level. The attraction values used were based on the average for facilities of that type within Gloucestershire. Jobs were added to the employment theme based on the number of people expected to be employed by each facility. Again this was based on the average employee size of similar facilities in Gloucestershire. Thus residents in those settlements were provided with access to a better range of facilities.

Table 15.6 Modelled annual travel resulting from Option 2 – a development strategy to tackle under provision in Gloucestershire.

Purpose	Mode	Trips per Capita per annum	Annual Travel per Capita (km)	Annual Energy per Capita (kg)	Mean Trip Distance (km/trip)
WORK	CAR	90.68	1,333	76.1	15.00
WORK	BUS	8.44	72	1.2	8.71
EDUCATION	CAR	32.78	238	14.8	7.40
EDUCATION	BUS	12.84	141	2.3	11.18
RETAIL	CAR	212.03	2,245	85.7	10.50
RETAIL	BUS	22.75	162	2.6	7.25
LEISURE	CAR	94.67	1,657	46.4	17.85
LEISURE	BUS	7.11	125	2.0	17.87
TOTAL		*486.32*	*5,986*	*231.0*	*12.50*

The model was rerun and the following results produced (Table 15.6). In general, this strategy resulted in a very slight reduction in average trip lengths. There was not much change for work trips but whether or not a settlement was under provided in terms of employment was not taken into account in the methodology for allocating services. Average trip lengths for leisure purposes have changed the most – reducing from 18.2 km per trip to 17.9 km per trip in the case of leisure trips by bus, a reduction of 1.75%. On the whole, bus trips seemed to be affected to a greater extent than car trips. This resulted in a very slight increase in the proportion of travel made by car – 91.59% in 1998, 91,63%. The effect on fuel usage was even less noticeable with a decrease of 0.2% in fuel used per km travelled and an increase of 0.2% in fuel used per trip. Spatially, this strategy resulted in decreased trip lengths around Stroud, Cirencester and the Forest Cluster (Figure 15.3), which were the main areas where facilities were added.

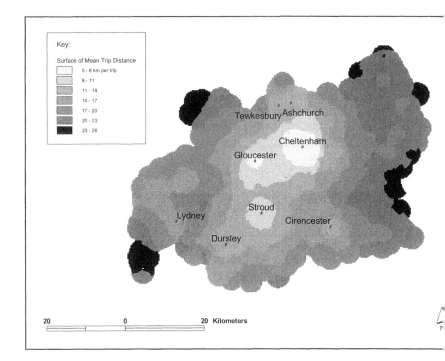

Key:

Surface of Mean Trip Distance

- 5 - 8 km per trip
- 8 - 11
- 11 - 14
- 14 - 17
- 17 - 20
- 20 - 23
- 23 - 26

20   0   20 Kilometers

Figure 15.3 The Pattern of Mean Trip Lengths Resulting From Option 2.

15.6.3 Option 3

In this strategy housing was allocated to those settlements which had been ide
having excess capacity in their levels of facility provision, *i.e.* were over pro
Population was added to each settlement so that none of the nine services inclu
study were then over provided. Characteristics of the population added, su
number in employment, number of school-aged children and car ownership le
based on the existing population of that area. In all, an extra 28,000 popula
added. Gloucestershire's forecast population for 2001 is 567,100 and fo
583,900, *i.e.* an increase of 26,900. So this method covers any projecti
additional facilities, services or employment were added as part of this allocatio

In general this created longer trip lengths as those settlements over p
terms of facilities tended to be smaller, more isolated settlements (Table 15.7)
adding population to balance the maximum difference, resulted in some faci
being under provided. Finally, the type of facility that tended to be over prov
pubs and garages, these don't necessarily serve the local community, *i.e.* d
passing trade in respect of pubs or are specialists, dealing with one type of car,
of garages. The percent of journeys made by car increased slightly (by 0.1%
levels), as did the distance travelled per capita (2%) and the fuel consumed
(2%). There were no discernable effects on the spatial distribution of mean trip

Table 15.7 Modelled annual travel resulting from Option 3 - a development strategy to tackle over provision in Gloucestershire.

Purpose	Mode	Trips per Capita per annum	Annual Travel per Capita (km)	Annual Energy per Capita (kg)	Mean Trip Distance (km/trip)
WORK	CAR	89.41	1,333	76.1	15.30
WORK	BUS	8.15	72	1.2	8.83
EDUCATION	CAR	32.16	238	14.8	7.44
EDUCATION	BUS	12.75	141	2.3	11.18
RETAIL	CAR	213.91	2,245	85.7	10.68
RETAIL	BUS	21.97	162	2.6	7.37
LEISURE	CAR	93.21	1,657	46.4	18.47
LEISURE	BUS	6.90	125	2.0	18.48
TOTAL		*478.46*	*5,986*	*231.0*	*12.79*

15.6.4 Option 4

In this strategy it was assumed that planning gains from developments would be used to provide additional services and facilities in locations that were under provided, whilst housing allocations would be used to support existing services and facilities in areas which are currently over provided. The method for implementing this strategy within the model was to combine the above two strategies, using origin data from the over provision strategy and facility data from the under provision strategy.

Table 15.8 Modelled annual travel resulting from Option 4 – a development strategy to tackle both under and over provision in Gloucestershire.

Purpose	Mode	Trips per Capita per annum	Annual Travel per Capita (km)	Annual Energy per Capita (kg)	Mean Trip Distance (km/trip)
WORK	CAR	91.13	1,396	79.5	15.32
WORK	BUS	8.30	73	1.2	8.77
EDUCATION	CAR	32.77	244	15.2	7.44
EDUCATION	BUS	12.99	145	2.3	11.19
RETAIL	CAR	218.01	2,323	88.9	10.65
RETAIL	BUS	22.39	164	2.6	7.31
LEISURE	CAR	95.00	1,740	48.5	18.31
LEISURE	BUS	7.03	128	2.0	18.17
TOTAL		*478.63*	*6,212*	*240.3*	*12.74*

This strategy produced slightly shorter mean trip distances than the strategy to tackle over provision of services (option 3) but still resulted in a slight increase (less than 1%) in mean trip lengths and a more substantial increase of 7% for both total distance travelled and fuel consumed compared with the situation in 1998 (Table 15.8). Fuel consumed per capita increased by 2% whilst there was no change in fuel consumed per km travelled or in the percentage of trips made by car as opposed to bus. As for the strategy for tackling under provision of services, this strategy resulted in shorter trip lengths for those living in Stroud, Cirencester and the Forest of Dean.

15.6.5 Option 5

The previous two strategies concentrated on placing additional population █ lower order services and facilities by adding small pockets of housing t█ settlements. This strategy takes a slightly different approach, concentrating ❑ supporting higher order services, particularly those that require a population in of 25,000 to survive.

In this scenario additional population was placed in such a way as to take of settlements above the threshold required to support higher order services currently provide. Research at the Bartlett showed that residents of settleme█ population of over 25,000 tend to travel shorter distances and use the car less █ in smaller settlements (Banister, 1999 and Williams, 1997). Williams (2█ identified a number of services and facilities that require a population of th█ support them. Thus, settlements within Gloucestershire that could be expanded in size were identified. Of the settlements under the threshold, two were clos█ level – Dursley and Cirencester. Housing was allocated to each of these settl█ bring them up to the required size. Key services were then added to the level █ be supported by the expanded settlements. The remaining housing alloc█ insufficient to bring a third settlement up to the 25,000 threshold, so was allocat█ smaller settlements to bring them above the 5,000 threshold, thus of a size t█ support services such as banks and supermarkets.

Table 15.9 Modelled annual travel resulting from Option 5 – a development strategy to aimed at su█ increased number of higher order services in Gloucestershire.

Purpose	Mode	Trips per Capita per annum	Annual Travel per Capita (km)	Annual Energy per Capita (kg)	Mean T Distan (km/tr█
WORK	CAR	90.81	1,325	76.0	1█
WORK	BUS	8.45	24	0.4	█
EDUCATION	CAR	32.58	231	14.5	█
EDUCATION	BUS	12.70	37	0.6	█
RETAIL	CAR	217.02	1,817	72.4	█
RETAIL	BUS	22.71	43	0.7	█
LEISURE	CAR	94.66	1,754	48.9	1█
LEISURE	BUS	7.11	78	1.3	1█
TOTAL		*486.05*	*5,310*	*214.6*	*1█*

This had considerable impact on the travel patterns of the county (Table 15.9). distance travelled by the residents of the county decreased by 9% over 1998 leve█ a 5% increase in the population of the county. Total fuel consumption dropp█ whilst average journey length decreased by 13%. These changes are very simila█ achieved through the structure plan strategy. The most dramatic decreases i█ lengths are for journeys by bus. This is partially due to the high densities assum█ the two major settlement expansions of Dursley and Cirencester and the inadec█ the model in coping with assigning trips to different modes under these condi█ reality it is likely that many of the very short bus journeys would have been mad█ motorised modes, thus an even more dramatic reduction in fuel consumption █ expected. Journey lengths in the areas surrounding all five expanded se█ decreased substantially (Figure 15.4). This was accompanied by a slight increas█

journey lengths in the area just to the northwest of Gloucester, with residents in these areas being attracted to the new facilities in Dursley and the smaller centres such as Micheldean which were expanded as part of this option.

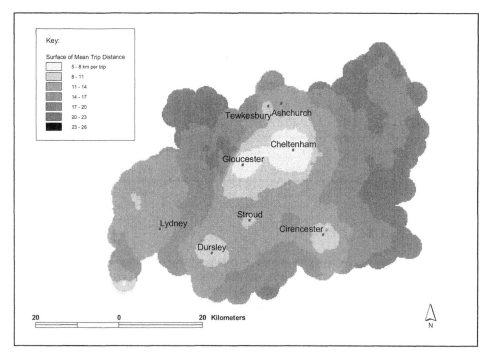

Figure 15.4 The Pattern of Mean Trip Lengths Resulting From Option 5.

15.6.6 Option 6

The final option tested takes a different approach. The basic premise was that housing could be allocated in such a way as to support public transport services. As no good quality data was available on the size of population needed to support particular types and frequencies of service the effect of an improved county-wide bus service on travel patterns of the current residents of Gloucestershire was modelled, rather than adding the population required for these changes to take place. This option assumed that all existing bus routes were run with a service frequency of a least once an hour, in line with findings that the proportion of trips made by bus increases dramatically if service frequencies are above this level (Titheridge, 2000).

The results showed a small but noticeable decrease in trip lengths (Table 15.10), distance travelled and energy used (1%, 2% and 1% respectively) compared with 1998 levels. The number of trips also decreases slightly as bus users are more likely to combine trips, so tend to make less journeys in total. The distance travelled per capita and fuel consumed per capita also decreased by 1% compared with 1998. More interestingly, this was the only strategy that resulted in a reduction in the proportion of trips made and distance travelled by car compared to bus.

Table 15.10 Modelled annual travel resulting from Option 6 – a transport strategy to tackle pc
transport provision in Gloucestershire.

Purpose	Mode	Trips per Capita per annum	Annual Travel per Capita (km)	Annual Energy per Capita (kg)	Mean Dista (km/t
WORK	CAR	89.00	1,327	75.9	
WORK	BUS	10.39	96	1.5	
EDUCATION	CAR	33.17	246	15.3	
EDUCATION	BUS	11.31	122	2.0	
RETAIL	CAR	211.14	2,213	84.8	1
RETAIL	BUS	27.22	210	3.4	
LEISURE	CAR	93.18	1,671	46.7	
LEISURE	BUS	8.29	157	2.5	
TOTAL		*483.70*	*6,042*	*232.2*	*l*

15.7 CONCLUSIONS

Only small changes in travel patterns were achieved through options 3 and
involved a housing allocation strategy that concentrated on lower order sei
facilities such as convenience stores and pubs (Table 15.11). Much greater
were achieved through option 5 – the strategy to expand as many towns as ʀ
above a threshold population size of 25,000. The travel reduction resul
expanding Dursley and Cirencester to this threshold was comparable with
reduction achieved through the structure plan (option 1). Thus, development
based around expanding smaller settlements to above a 25,000 population thresl
have considerable impact on travel reduction. This conclusion is support
findings of Banister (1999) and Williams (1997) from analysis of national tra
data for different settlement sizes that residents of settlements with a populatiؑ
25,000 tend to travel shorter distances and use the car less than those ꞽ
settlements. Allocating housing in this way could not only reduce travel
significantly increase the level of access to key services and facilities
populations.

The travel reduction resulting from option 5 was comparable with
reduction achieved through the structure plan (option 1). It should be remem
the technique was applied to only a very limited range of facilities and employ
were not considered. If a wider range of services, facilities and employment
included in the analysis it is likely that much greater reductions in travel could
achieved.

The results of the modelling exercise also suggest that the key to moving ʈ
more sustainable modes is to accompany any new development with improv
transport services. Consideration needs to be given to the size of populatio
support frequent bus services. The minimum population levels could be p
obtained through development along public transport corridors linking major ʈ
ring of smaller settlements served by a circular bus route.

Table 15.11 Summary of changes in travel resulting from different development strategies.

	Population	Travel per capita* (passenger-km)	Mean Trip Length* (km/trip)	Modal Split[1] (% passenger-km)	Fuel per capita (kg)	Other Benefits
Base (1998)	557,000	5,986	12.56	Car – 89.5% Bus – 10.5%		
			Change from Base			
Option 1	27,000	-13.20%	-13%	+CAR	-8.40%	Low development costs as reduced need to provide extra facilities
Option 2	As Base	-0.40%	0%	+CAR	-0.20%	Improved access to facilities
Option 3	28,000	2.20%	2%	+CAR	1.90%	Lower development costs as reduces need to provide extra facilities
Option 4	28,000	1.80%	1%	+CAR	1.70%	Improved access to facilities
Option 5	27,000	-13.00%	-13%	+CAR	-9.20%	Improved access to high order facilities
Option 6	As Base	-1.10%	-1%	+BUS	-1.70%	Improved access to facilities for non-car users

*Includes internal trips only.

It is recognised that expanding all existing settlements to a population of 25,000 is impractical. Whilst the resulting changes from allocating additional housing to support lower order services and facilities were small, it is felt there are additional benefits to be derived from allocating housing to support services and facilities in smaller settlements. Also, the methods used for identifying areas of under and over provision and then for balancing population with facility provision were crude, possibly effecting the results that could be achieved in travel reduction through assigning development based on current facility provision. Consequently, it is felt that this technique shows potential for being a useful tool for prioritising developments.

Several ways in which the methodology for assigning housing locations suggested in this paper could be improved have been identified. Clearer guidelines are needed when identifying suitable sites for development. Some locations are clearly inappropriate for housing development in terms of access to higher-order facilities, lack of public transport provision *etc.* It was also found that no settlement was consistently over provided for in all facility types. Priority needs to given when allocating housing to supporting certain facility types, more work is needed to identify which facilities should be prioritised. Finally, the methodology needs to be expanded to include wider range of facilities and some measure of facility quality.

15.8 REFERENCES

Action for Communities in Rural England, 1998, *ACRE's Detailed Commer Comprehensive Spending Review for the Department of the Environment, and the Regions.* (Cirencester: ACRE).

Banister, D. 1999, Planning more to travel less: land use and transport. *Town Review,* **703**, pp. 331-338.

Council for the Protection of Rural England, 1998, *Rural Transport Policy ar* (London: CPRE, Royal Development Commission and The Cc Commission).

Department of the Environment, Transport and the Regions, 1997, *Nation Survey 1994/96.* (London: The Stationary Office).

Department of the Environment, Transport and the Regions, 1999, *Proje Households in England to 2021.* (London: HMSO).

Department of the Environment, Transport and the Regions, 2000a, *Plannii Guidance Note 3: Housing Revised.* (London: HMSO).

Department of the Environment, Transport and the Regions 2000b, *Nation Survey 1997/99 Update.* (London: The Stationary Office).

Forest of Dean, 1996, *Local Plan Adopted Version.* (Coleford: Forest of Dea Council).

Gloucestershire County Council, 1999, *Gloucestershire Structure Plan* Adopte November 1999.

OPCS, 1994, *1991 Census Workplace and Transport to Work: Great Britain* Vol.1. (London: HMSO).

RTPI, 1998, *1995 GIS Survey.* (London: RTPI).

RTPI, 2000, *IT in Local Planning Authorities 2000.* (London: RTPI).

Titheridge, H. 2000, Balancing housing and facility provision: the transport imp *URBASSS Working Paper 8*, Sept. 2000. (London: The Bartlett School of University College London).

Titheridge, H. and Rana, S. 2000, ESTEEM: a technical report, *ESTEEM Work 8*, June 2000. (London: The Bartlett School Of Planning, UCL).

Titheridge, H., Hall, S. and Gardner, R. 1998, Sustainable settlements – a moc estimation of transport, energy and emissions: model development and ca *ESTEEM Working Paper 3*, Oct. (London: The Bartlett School of Planning,

Titheridge H, Hall S And Banister D 2000, Assessing The Sustainability Development Policies In *Achieving Sustainable Urban Form* edited by Wi Burton, E. and Jenks, M., (London: E & FN Spon), pp. 149-159

UWE, 2000, *The Interdependence between Urban and Rural Areas in the England.* (Bristol: University of West of England), June 2000.

Williams, J. 1997, A study of the relationship between settlement size and trave in the UK. *URBASSS Working Paper 2*. (London: The Bartlett School Of University College London).

Williams, J. 2000, Tools for Achieving Sustainable Housing Strategies Gloucestershire. *Planning Practice and Research*, **15**(3), pp. 155-174.

PART V

GIS in Socio-Economic Policy

16

Using GIS for sub-ward measures of urban deprivation in Brent, England

Richard Harris and Martin Frost

16.1 INTRODUCTION

Recently there has been increased interest in defining and locating areas of poverty, deprivation and social exclusion in the UK. Such terms are difficult to define in any precise and apolitical sense. Nevertheless, new measures have been devised that aim to calculate deprivation and poverty rates in consistent, robust and (pseudo-) scientific ways. These measures include the Department of the Environment, Transport and the Regions' (DETR) Index of Multiple Deprivation – IMD 2000 – and also the Poverty and Social Exclusion Survey of Britain (DETR, 2000b, Gordon *et al.*, 2000). The IMD 2000 statistics are easily accessed from the National Statistics Service (NSS), a website developed by the Office of National Statistics (ONS), in partnership with central and local government (see www.statistics.gov.uk). The aim of the NSS is to make statistical information available for small areas across the UK. Presently the service offers statistics at only the Ward (electoral district) level. However, the intention is to introduce smaller geographical units based on the 2001 UK Census Output Areas. Average household income estimates will then be assigned to those units. On the basis that better information begets better policy-making – a rationale behind the NSS – then this is an important development. This chapter highlights the need for geographically meaningful income estimates, based on flexible approaches to model building and area classification.

According to ONS (2000), one of the benefits of the NSS will be to assist neighbourhood renewal; "to facilitate a better understanding of local problems and effective targeting of solutions" by allowing assessment of local need. It should therefore prove invaluable to the various policy-makers and agencies that have responsibility to counter the effects of rising proportions of deprived households within the UK (14% of households in 1983; 20% in 1990; 24% in 1999: Gordon *et al.*, 2000). At a national scale, income inequality has also risen, reaching its highest level since 1981 (when comparable data were first collected). The 1999/2000 annual Family Expenditure Survey (ONS, 2001) shows that the poorest fifth of households had 6% of national income after tax, while the share held by the top fifth has risen under the Labour government, from 44% to 45% (Ward, 2001) – not a statistically significant rise, but one that nevertheless reveals the difficulties of reversing the legacies of income inequality in the UK.

In our view a greater sensitivity to geographical difference is required to really begin unlocking the true picture of deprivation, inequality and social exclusion across

both the UK and within specific regions. A social paradox of current tim*
processes of both 'ghetto-isation' and fragmentation appear to exist, and
proximity (Urban Task Force, 1999, Hall and Pfeiffer, 2000). The consequer
whereas some neighbourhoods will contain a large proportion of deprived h
others will also contain 'hidden' deprivation, localised at a more niche scale.
urban White Paper to be published by a British government for some tw
recognises that "even in those towns and cities with significant deprivation ther
a sharp contrast between prosperous areas and those with most deprivation. For
Sheffield has two wards amongst the least deprived in the country just acros
from some deeply deprived areas" (DETR, 2000a). Despite this insight and its
on wards, ward boundaries were actually designed for administrative purposes *
monitoring urban deprivation or for targeting neighbourhood renewal *per se*.
little reason to suppose that geographies of deprivation are adequately describe*
dividing lines that happen to coincide with the boundaries of the UK's electoral g

Areal indicators such as IMD 2000 are suited to identifying neighbou
deprivation at a ward scale only but are insufficient for either identifying h
living in poverty at a sub-ward scale, or for identifying flows of deprivati
administrative boundaries. They are designed to enable between ward compa
without knowledge of within-ward heterogeneity. It is implicitly assumed t
provide a suitable foundation on which to build spatial comparison. That is an a
which we contest, using two additional datasets to highlight how deprivation
within wards and also cross ward boundaries (a typical ward contains, on aver
2000 households, although individual wards vary considerably in both their p
and physical size). The first data source is based on local administrative record
London Borough of Brent that allow identification of households receiving ke
Using a simple, point-pattern analysis undertaken using analytical tools commo
GIS packages we identify local concentrations of households whose children re
school meals. We then compare the distribution of those households against th
deprivation domain of IMD 2000 (DETR, 2000a). The second source of inform
commercial, lifestyles dataset collected from individual respondents to a natior
but tabulated at a unit postcode level (for reasons of privacy and data protectic
UK, a unit postcode contains an average of 12 to 16 residential, delive
(letterboxes). Using income data taken from this dataset we again undertak
pattern analysis, identifying, within the Brent study region, local concentratio
income households. The geography of low income revealed by the lifesty
analysis is also compared against the IMD 2000. Our analysis suggests that in
in GIS, and in data collection and data handling facilitate the use of 'uncon
sources of data to at least complement ward-based deprivation indicators. The
is more on that contention than on the detail of the rather UK specific datasets e

16.2 DEPRIVATION, THE INCOME DOMAIN AND BRENT

Deprivation is by no means an exclusively urban phenomenon, and, of cours
urban areas are deprived. Yet, based on the DETR (2000b) statistics, over 7(
population who live in one of the 10% most deprived wards (boundaries as at
1998), also live in one of the main conurbations: Greater London; Greater M
Merseyside; South Yorkshire; Tyne and Wear; West Midlands; West Yorkshi
former county of Cleveland (DETR, 2000a). DETR (2000a) reports that peopl
the English conurbations have on average, and by comparison to the populatio

lower educational results; lower employment rates; more children living in poverty; an higher mortality rate; and an increased exposure to violent crime and theft.

The IMD 2000 income domain – which we now refer to as simply 'the income domain' – is used to monitor income deprivation by identifying low income families from Department of Social Security benefits data. It is one of six domains that are combined, with weighting, to form the composite IMD 2000. The full set of domains and their weightings are: income (25%); employment (25%); health deprivation and disability (15%); education, skills and training (15%); geographical access to services (10%); and housing (10%). Both the income and employment domain scores are calculated as rates. Consequently, if ward X has a score of 40% on the income domain, then it is estimated to have twice as much income deprivation as ward Y with a score of 20%. In our analysis we consider only the income domain as, intuitively, this ought to be the most closely associated with the other income information we have available to us. Further details about IMD 2000 are available in DETR (2000b) or at http://www.regeneration.detr. gov.uk/research/id2000/index.htm.

We have chosen Brent as a study region because it is indicative of a relatively small urban area containing a diverse mix of social, economic and demographic conditions. For instance, the housing stock has fragmented into a complex mix of tenures and despite some redevelopment (*e.g.* of 1960s/70s tower blocks) there remain problems of 'suburban decay' and a diminishing local economy. At the time of writing the largest disused building within the Borough is probably Wembley Stadium! Figure 16.1 shows the income domain scores for each of the 31 wards within Brent. The scores range from 53% of the population living in income deprived families in Stonebridge, to 12% of the population in Keniton. For comparison, the most income deprived ward in England has 74% of its population living in low income families. That is a 40% increase over Stonebridge but is exceptional. Stonebridge ranks as the 111th most deprived ward in England (of 8,414), meaning it is measured to be within the top two percent most income deprived wards nationally. In fact, four Brent wards are within the top two percent nationally, and a further three are within the top ten percent. None of the Brent wards are in the least deprived quartile nationally. DETR (2000b) reports that there is a total of 72,381 people who are income deprived in Brent.

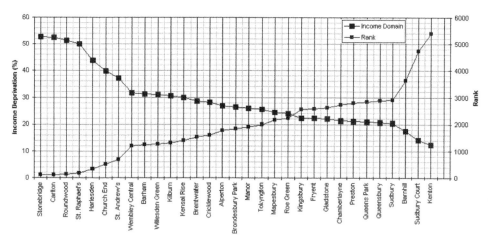

Figure 16.1 Income deprivation measured for Brent wards.

Figure 16.2 maps the data shown in Figure 16.1. It is a choropleth map
income domain scores organised into quartiles – four classes with an (appro
equal number of wards in each (but not necessarily equal population). As with
based on the UK census or electoral geography, it is not a realistic representa
urban morphology insofar as each area contains an undisclosed mixture of r
commercial, derelict and open land (plus some areas of water). Harris and
(2000) suggest using Ordnance Survey UK's Code-Point product as a basis
filling out from unit postcode centroids (which are listed in the dataset with a
precision in the majority of cases), using surface estimation and modelling pro
identify areas of either residential or commercial land-use. Here we are int
locating residents (who are deprived) and so our map would provide better
information if we excluded non-residential areas. To do so we have
sophisticated techniques than those adopted by Harris and Longley (2000), but
are essentially standard to any GIS package: we first created a 100 metre b
around each of the properties defined as residential within Brent authority
property database (the database was recently updated and ascribes a geo-refere
metre precision to every property listed within it – ideally, every property in B
then dissolved the buffer zones into one theme (or map layer); finally, we use
theme as a template to 'cookie cut' out the residential areas of Brent from
residential, census geography. The resulting map layer is shown in Figure 16.3
property locations also shown) and is, of course, an artefact of the modelling pr
specifically, the source data and the buffer width. That buffer width, of 100
somewhat arbitrary but was selected to ensure that buffers drawn around
points overlapped, meaning they can be merged (dissolved) together to prod
that does not appear overly fragmented in terms of its residential area. A
generalization of the true residential limits, Figure 16.3 is still a more realistic
Brent's residential geography than the census geography of Figure 16.2.

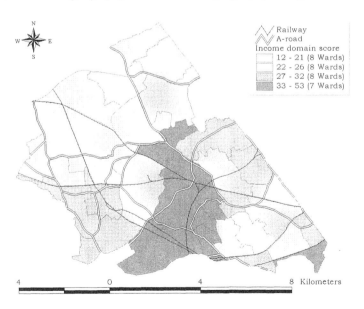

Figure 16.2 Income domain scores for Brent wards.

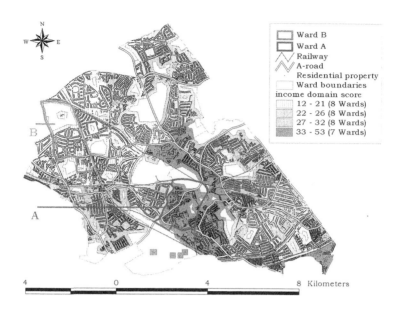

Legend:
- Ward B
- Ward A
- Railway
- A-road
- Residential property
- Ward boundaries

income domain score
- 12 - 21 (8 Wards)
- 22 - 26 (8 Wards)
- 27 - 32 (8 Wards)
- 33 - 53 (7 Wards)

4   0   4   8 Kilometers

Figure 16.3 Residential areas in Brent.

16.3 FREE SCHOOL MEAL ELIGIBILITY AS A MEASURE OF DEPRIVATION

Two wards are highlighted in Figure 16.3, and labelled as A and B. Ward A has an income domain score in the upper quartile for Brent; ward B a score in the lower quartile. Since it is low income households who are most eligible to receive free school meals, so we would expect to find a greater concentration of free meal households in ward A, than in ward B. A cursory inspection of Figure 16.4 – which maps the geographical distribution of free meal take-up within the study region – suggests the proposition to be true: there appears to be a greater number of free meal recipients in ward A than in ward B.

The centre of each circle plotted in Figure 16.4 is defined by a geo-reference assigned from the Brent property database to an household receiving free school meals. Approximately 80% of free school meal recipients listed in the benefits database have been matched, by address, to the property database. This ascribes a precise location to each household but, given that this is personal information, requires permission to be obtained from the UK Data Protection Registrar for data storage and analysis. The size and shading of each circle indicates the local concentration of free school meal households around and including each single household point. In detail, a simple point-pattern analysis has been undertaken that operates by centring a circular window, of radius 500 metres, on each of the 1,634 household points that receives free school meals. The total number of recipient households within the 500 metre radius focal region is then found and that value assigned back to the central point (the household record). The outcome of this procedure is one of aggregation, drawing-out local trends in the data, whilst, in principle, partially smoothing out random errors. The procedure is analogous to operating a low pass filter on remotely sensed imagery. Again, interpretation of the results is contingent on the modelling procedure; specifically, the somewhat arbitrary radius specification, subsequent classification of the results (quartiles in Figure 16.4) and

also the map symbols (varying the size and the density of their shading can em
de-emphasise apparent geographical trends). Note that because the circu
windows overlap across the study region, so the sum of the modified data valu
greater than the original sum of 1,634 households. The effect is one of double
(or more). If the circles did not overlap then only the central point would ever b
each focal window and, consequently, the data values would not be changed
metre radius has been selected to ensure overlap and because, from experience
a suitable balance that neither over-, nor under-smooths the data. However, it
noted that our implicit assumption that the 500 metre value is 'correct' is a
unproven assertion which has not, in the context of this chapter, been subjec
rigorous testing.

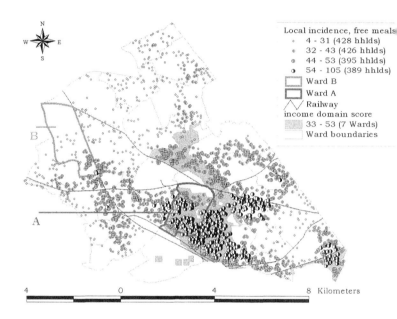

Figure 16.4 Simple point-pattern analysis of free school meal households.

In Figure 16.4, households that receive free school meals appear geog
concentrated in wards with relatively low income domain scores. Such house
not exclusively contained in the most income deprived wards, however. Table 1
the distribution of free meal households across the quartile grouping of Brent
income domain score. The table shows that whilst there is an higher inciden
meals allocated to households in the poorest quartile wards (516 household
share), over two-thirds of households receiving free school meals are within oth
indeed, 315 (19%) are in the relatively most affluent wards.

In the analysis we are not actually comparing like with like, but an absolut
(number of free meal households) against a rate (income domain score). It cou
the poorest quartile of wards contains the highest number of free meal househol
because it also contains the largest share of the total population (there being no
that 1991 Census electoral wards in any part of the UK will have near equal p
sizes). In other words, a larger number of people would, *ceteris paribus*, lead to

number of eligible households. As it happens, the observed share of free meal households for the poorest quartile *is* (at 32%) greater than the expected share from the total population alone (27%). Conversely, the observed share in the most affluent quartile (19%) is less than the expected share from the population (23%) – see Table 16.1. However, the observed and expected values are also positively correlated: a 0.60 correlation using Spearman's rank statistic; 0.61 with Perason's correlation coefficient; and a 0.55 likelihood that the observed and expected values are not independent, based on a chi-squared analysis. Caution should be applied when interpreting these statistics since the magnitude of any correlation tends to increase with aggregation and here we are considering a broad grouping of wards into only four classes. Furthermore, free school meal up-take is a narrow and unreliable measure of income deprivation that necessarily excludes families without children of school age and the more aged members of society. It is also not perhaps the case that take-up of free school meals correlates well with eligibility to that right. What we can say with certainty, however, is that there are at least 315 households that are income deprived (insofar as they receive free meals) but who are also living in relatively affluent wards.

Imagine the results reflect a general trend: that one-fifth of all income deprived households are living in wards that are not classified as particularly deprived on the income domain scale. This would raise obvious concerns about the use of broad scale, area measures for redistributing remedial funds to the most needy. Of course, we cannot and do not substantiate such concerns based our results alone. The one fifth value should rightly be considered speculative. Nevertheless, what we do argue, and can show, is that pockets of deprivation exist at sub-ward scales but these are 'hidden' or 'averaged-out' by ward scale indicators that have no corresponding measures of internal variation or diversity (and to define a mean without also defining the variance is to only tell half the story). That contention is a prelude to the next section where we use simple GIS techniques to search for 'niche' pockets of deprivation on the basis of information taken from a commercial dataset that was originally complied primarily for the purposes of direct marketing.

Table 16.1 Distribution of free school meal households across quartile grouping of Brent wards by income domain score.

	Number of free meals	Share of total (%)	Share of all households (%)
Quartile 1[*]	315	19	23
Quartile 2	389	24	22
Quartile 3	414	25	28
Quartile 4[+]	516	32	27

[*]Quartile 1 has least income deprivation.
[+]Quartile 4 has most income deprivation.

16.4 TARGETING CLUSTERS OF DEPRIVATION WITH LIFESTYLES DATA

Figure 16.5 shows estimates of the proportion of households in each of 11 income bands for the two wards named previously as A and B. The income bands increase at £5,000 intervals from zero, until reaching the eleventh band which is open-ended (total family

income of £50,000 or more, per year). The data source is a commercial,
database formed by replies to a national, consumer survey. That survey ⬤
undertaken in England, Wales and Scotland by a commercial data vendor
spring of 1999, and sent out to a high proportion of households for whom a
member was listed on the Electoral Register for the three countries. (In princip⬤
should be listed on the Register since it is a legal offence not to be so. Howe⬤
somewhat of an ideal). The full lifestyles dataset contains a wide range of socio⬤
behavioural and consumer information that describe approximately o⬤
households. From that dataset we have extracted 1,825 records corresp⬤
households within Brent who responded to the survey question 'which ⬤
describes your *combined household income*?' (survey's original emphasis).
Mail's Postal Address File (PAF) for the same year lists 98,314 residential, ma⬤
points in the Borough. On that basis the lifestyles sample is of just un⬤
households within Brent, which is consistent with the national average. Within
sample is of 2.4% and in ward B, 2.3%. Strictly speaking these are slight ove⬤
since a single mail delivery-point is sometimes shared by two or more hous⬤
example converted terraced properties where a single letterbox serves tw⬤
apartments).

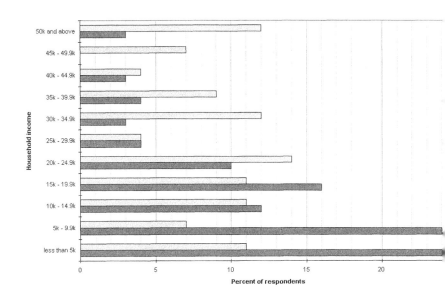

Figure 16.5 Household incomes in two Brent wards.

Figure 16.5 reveals that in ward A, a minimum of 10 of the 11 income g⬤
be found. We know this because the lifestyles survey asked real people about t⬤
income status. Unless the respondents lied or were mistaken, then we know
least one household present from each of ten income groups. In ward B, all 11 ⬤
present. The results point to diversity at sub-ward scales, but not chaos. Figure⬤
suggests evidence of geographical difference, with ward A containing higher p⬤
of lower income families (notably those earning under £10,000 per year). Alt⬤

finding is entirely consistent with ward A having a lower IMD 2000 income domain score, we are *not* claiming that use of lifestyles data will necessarily yield precise and accurate estimation of income deprivation rates at ward or sub-ward scales. To the contrary, Longley and Harris (1999) have previously discussed the problems associated with using lifestyles data in socio-economic research. Although some of their findings are specific to a different dataset, their general conclusions remain valid here. In particular, the survey respondents are recognised as essentially self-selecting, deciding whether or not to return a survey questionnaire, and, as such, they may not form a representative sample of the population-at-large.

The self-selection of survey respondents should not be regarded as solely a random error. For example, Harris (1999) gives limited evidence that it is the 18-24 age group, and also the most and least affluent members of society who are least likely to respond to a consumer survey of the type considered here. Yet, neither can the self-selection easily be treated as giving rise to a systematic bias which could, in principle, be corrected by weighting the data against a second (more accurate) dataset. The problem here is twofold. Firstly, there is no obvious dataset that can be used to 'ground truth' the lifestyles data other, perhaps, than at comparatively coarse scales of aggregation (where official social survey data might be used). Secondly, the propensity to respond to the consumer survey is ultimately a matter of individual choice and is not unambiguously related to any 'obvious' variables like age, affluence, location or lifestage. The propensity to respond retains a random component. Furthermore, though there is, arguably, a decreased response from the less affluent, that is not to say no lower income households will respond (as Figure 16.5 proves). With all this in mind, what we suggest is that the lifestyles data can be used to identify, at the unit postcode level, the location of every survey respondent who indicated their annual family income to be less than £10,000. If these locations can be shown to be clustered spatially, then we would have cause to believe that they are likely areas of income deprivation, particularly if a reasonable number of the clusters were shown to be within, or in proximity to wards with high income deprivation according to the IMD domain scores.

In accordance with our argument, Figure 16.6 maps the location of households with annual family income less than £10,000 who responded to the lifestyles survey. The point-pattern methodology described in Section 16.3 has again been used to draw-out patterns in the data. A difference is that the centre of each circle is now defined by the location of a unit postcode centroid, ascribed to each household from the Postal Address File (PAF) with a precision of 100 metres or greater. In general there is a large degree of correspondence between the geography of income deprivation revealed from the lifestyles analysis and that suggested by the income domain scores. A visual inspection of Figure 16.6 shows that most clusters of low income households are either fully or partly within wards that are in the upper quartile of income domain scores for Brent (the most income deprived). Yet, not all are. Furthermore, there are a number of instances where the ward boundaries are revealed to artificially partition low income clusters into two or more different areas. In relation to cartography and its application in policy analysis, Figure 16.6 is, perhaps, a more honest map than Figure 16.2, say, because Figure 16.6 gives a more vivid impression of where income deprivation is to be found in wards and also allows the reader to gain an impression, albeit partial, of social heterogeneity within the region. The uniformity of populations that census-based mapping implies is not assumed. In particular, Figure 16.6 makes clear a central message of this report: income deprivation does not respect ward boundaries!

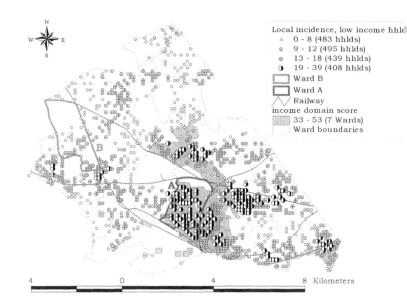

Figure 16.6 Point-pattern analysis of lifestyles data to identify low income households.

16.5 CONCLUSION: A FLEXIBLE APPROACH FOR AREA CLASSIFI*

Whilst we welcome the development of the National Statistical Service and
access it offers to policy relevant data, our report has shown how too rapid (an
glib) identification of 'poor areas' within cities can conceal a dispersion
deprived households in areas that are otherwise regarded as relatively affluent *o*
The intention of the NSS to provide smaller scale income estimates based
output areas will be an important step forward for neighbourhood analysis.
average measures should be set alongside measures of income variation, at the s*
An average statistic is never truly meaningful until a measure of diversity (v
also provided.

We have sought to develop the basis of a simple, 'bottom-up' ap*
identifying income deprived areas, using analytical tools available in ArcGIS
Harris, 2001). A refinement to the methodology would be to replace the fixed
the point-pattern analysis with an adaptive kernal that is more sensitive to the *
geography and urban morphology surrounding each point. Measures of
significance could be introduced, identifying whether a low income cluster*
'unusual' or merely, perhaps, an artefact of residential postcode geograp*
uncertainty of the lifestyles data could also be more explicitly incorporated *
modelling procedure. Such methods exist, having been developed in the
population surface modelling (Martin, 1998) and geocomputation (Longley, B*
McDonnell, 1998, Caldwell, 2000: see also http://www.geog.leeds.ac.uk/*
ccg.html). Here, however, we are more immediately concerned with applying
are available to users of 'standard GIS'.

Our report has suggested that it is possible to experiment with non-official or non-governmental sources of data, within a GIS framework, at sub-ward levels of analysis. The application of lifestyles data, that are regularly up-dated by annual survey, may provide a useful component of the suggested bottom-up approach, if – and it is a 'big if' – issues of representation and survey/response bias are suitably resolved. We provisionally suggest that the data can be aggregated in geographically sensitive ways to a level where use of the data is statistically robust and where the scale of analysis is not overly coarse. There is no reason why the aggregation should be bounded by geographically inappropriate administrative units and, by the act of aggregation, some of the ethical and confidentiality concerns of using detailed, household data are avoided. The possibility is opened-up of identifying, at small spatial scales, subtle dimensions of poverty and exclusion to complement the accelerating development of standard neighbourhood statistics. It is common practice in standard geodemographic or area profiling to group areas according to their socio-economic 'type', then using a range of descriptive statistics to identify the differences between the various classes. A similar approach could be adopted here, using the lifestyles data to search for different components of deprivation that distinguish the geographical clusters of, say, low income households. For example, there is evidence of an age related dimension within Brent, with the smaller clusters to the north west of the region (near ward B) having a greater incidence of widowers and people wearing an hearing aid than the larger south-central cluster (near ward A).

These are avenues for future research, however. For the present, we note that, for the first time, some lifestyles data have been made directly available to academics who register at MIMAS (`http://www.mimas.ac.uk/docs/experian/`). Although the data are currently limited to age, vehicle, property, household and population records at the postal sector level, the information is up-to-date and will be of interest to a number of researchers.

16.6 ACKNOWLEDGEMENTS

We are grateful to Brent local authority and to Claritas Europe for the supply of the data analysed in this report. The lifestyles data are copyright © Claritas Europe (`http://www.claritas.com`) and are reproduced with permission. The responsibility for any errors or omissions arising from the analysis are solely our own.

16.7 REFERENCES

Caldwell, D.R., 2000, Editorial: developments in geocomputation. *Computers, Environment and Urban Systems*, **24**, pp. 379–382.

DETR, 2000a, *Our Towns and Cities: the Future: Delivering an Urban Renaissance.* (London: The Stationary Office).

DETR, 2000b, *Indices of Deprivation 2000.* `http://www.regeneration.detr.gov.uk/rs/03100/index.htm`.

Gordon, D., Adelman, L., Ashworth, K., Bradshaw, J., Levitas, R., Middleton, S., Pantazis, C., Patsios, D., Payne, S., Townsend, P., Williams, J., 2000, *Poverty and Social Exclusion in Britain.* (York: Joseph Rowntree Foundation).

Hall, P., Pfeiffer, U., 2000, *Urban Future 21.* (London: E & FN Spon).

Harris, R.J., 1999, *Geodemographics and the Analysis of Urban Lifestyles.* U
Ph.D. Thesis (University of Bristol: School of Geographical Sciences).

Harris, R., 2001, On the diversity of diversity; is there still a place for
classification? *Area*, **33**, pp. 329–336.

Harris, R.J., Longley, P.A., 2000, New data and approaches for urban analysis:
residential densities. *Transactions in GIS*, **4**, pp. 217–234.

Longley, P. A., Brooks, S. M., McDonnell, R., (editors) 1998, *Geocomputation.*
(Chichester: John Wiley & Sons).

Longley, P.A., Harris, R.J., 1999, Towards a new digital data infrastructure
analysis and modelling. *Environment and Planning B*, **26**, pp. 855–78.

Martin, D., 1998, Automatic neighbourhood identification from population
Computers, Environment and Urban Systems, **22**, pp. 107–120.

ONS, 2000, Internet document. http://www.statistics.gov.uk/neigh
downloads/nssrev5.pdf.

ONS, 2001, Briefing paper. http://www.statistics.gov.uk/pdfdir/etbrief

Urban Task Force, 1999, *Towards an Urban Renaissance.* (London: E & FN Sp

Ward, L., 2001, Earnings inequality widens under Labour. *The Guardian*, A
http://www.guardian.co.uk/Archive/Article/0,4273,4171558,00.html

17

The spatial analysis of UK local electoral behaviour: turnout in a Bristol ward

Scott Orford and Andrew Schuman

17.1 INTRODUCTION

There is growing interest in the factors that influence turnout in elections in the UK. This concern has become paramount with the extremely low turnout in the 2001 General Election. Although research in this area is well established, there has been very little concerned with the geography of turnout, particularly at the local level. This research aims to address this omission by examining the geographical factors that influence turnout in a local election in a ward in Bristol, UK. A GIS of the ward was constructed using voting data taken from the marked-up electoral register used in the local election. The results suggest that both contextual factors, such as the size of the household in which the voter lives, and geographic factors, such as the distance from the household to the polling station, are important in understanding the propensity to vote.

17.2 BACKGROUND

Studying election turnout has long been a useful way of examining both democratic and societal participation. Elections are frequent, occur at different spatial scales, and their results are easily quantifiable. Recent trends in the UK show generally falling levels of turnout by voters in elections. Although not universally acknowledged as problematic, most academics and politicians see declining turnout as symptomatic of a general political and social malaise, which produces a democratic deficit, apathy and indifference to societal issues. Both the EU and British Government have not only expressed their concern, but also provided funds for substantial research in this area, through for example the 5th Framework RTD programme and the ESRC's 'Democracy and Participation' programme. Similarly at the local level, many local authorities in the UK (for example, Bristol) have established 'Democracy Commissions' tasked with examining the problems of turnout and participation in localities.

17.3 ELECTORAL GEOGRAPHY

Geography remains important in all election studies for a number of reas·
elections are organised geographically, through defined constituencies ai
Second, election results often show distinctive geographical variations in votin;
the most well known in the UK being the North-South divide of the 1980s ¹
generally Conservative-voting South and a Labour-voting North (see Johns·
1988). Third, voting is always place-specific: local factors and political atti·
always affect voting decisions. Fourth, electoral representation (seats rather thai
also distinctively geographical – for example in the 1997 General Ele·
Conservative party gained 17.5 % of the votes in Scotland but gained no
contrast, the Liberal Democrats with only 13 % of the votes gained 10 out of the
72 seats – primarily because, unlike the Conservatives, they got their votes ii
places. Lastly, and following on, there is a geography to the power and policy t·
from such a political representation.

The reasons for low electoral turnout can be broadly categorised into soc·
exclusion, alienation), social/administrative (political institutions, structures an·
mobilisation) and administrative (voter facilitation). While evaluating ways of
electoral participation has been much discussed (Miller, 1988; Rallings and
1990, 1994, 1997; Rallings *et al.* 1996; Rallings *et al.* 1994), little of this wo·
fully taken on the embeddedness of all participatory process, social and admi·
posited by Agnew (1987, 1996) and demonstrated by Schuman (1999).
overlooked in most voting studies is the particular influence of local factors o·
attitudes and voting decisions.

Since Cox's (1969) seminal work on the importance of the local ge·
context to voting, much has been studied and developed both with specific i
voting (Agnew, 1987, 1996) and broader regard to place and the contextuality·
(Johnston, 1991; Thrift 1983). Whilst there remain sceptics (Rose and McAllis·
McAllister and Studlar, 1992) the relationship between voting activity and
activity takes place has been increasingly demonstrated in a number of areas. S·
have been seen between voting and spatial variations in economic prosperity (
Johnston, 1995; Pattie *et al.* 1997) as well as between voting and local conditio·
local campaigning (Denver and Hands, 1997; Pattie *et al.* 1995; Schuman, 1999·

Less demonstrated however (with the exception of Zuckerman *et al.* 1994·
the 'neighbourhood effect' which suggests that an important factor in local con·
social networks that the individual voter participates in. Only recently has this ¡
to be remedied and Pattie and Johnston (1999, 2000) take some key steps·
Despite an extensive survey of existing work on the neighbourhood effect (see a·
and Johnston, 1979; Books and Prysby, 1991, 1999; Huckfeldt and Sprag·
Miller, 1977) they bemoan the fact that few studies have investigated the hypo·
"people who talk together vote together" directly rather than through inference·
the work to date, they argue, fails both to uncover the mechanisms b·
neighbourhood effect, and fails to deal with it at an appropriate scale, far too of·
on large scale constituency or regional data.

Using two hundred and eighteen polling districts as 'sampling points', .
Johnston (2000) conducted an investigation into party conversational effects·
1992 British Electoral Survey which had included questions about whom peopl·
about politics. While there were differences in the strengths of the
conversational/political effects between parties (usually because of other ·
effects like variations in party campaigning intensity), they provide clear evi·
conversation and context influence voting. Dividing the conversation effec·

'family' (spouses and other relatives) and 'non-family' (work, friends, neighbours *etc.*) a further distinction is noted: "...people who spoke with their kin were 4.9 times more likely to switch their support to the Conservatives" (pp. 59) whereas with non-family they were only 2.5 times more likely to switch. Similarly for Labour, the figures were 5.6 times for family and only 2.2 times for non-family, a huge family bias. Their results strongly suggest that "people listen most attentively, and are converted by, discussants from within their families. Non-family discussants are also influential, but not as strongly: the main sway occurs as a result of within-family discussions." (pp. 59).

In providing a challenge to the assumption of methodological individualism that underlies many British voting studies Pattie and Johnston (2000) suggest a number of avenues. The first concerns the neighbourhood effect itself. Pattie and Johnston (and others too) concentrate on the *party* neighbourhood effect, looking at the way in which people socialise each other to vote for a particular party. This will always have methodological problems in that voting is by secret ballot. If, rather, we examine the *voting* neighbourhood effect (*i.e.* people socialising each other to vote, not just for a party but rather to vote at all (and the two are obviously related)), then there are a number of methodological opportunities, explored below, as well as many potential policy applications.

17.4 CONTEXTUALISING VOTING BEHAVIOUR

In the tradition of place-based research (Holt and Turner, 1968; Milne and McKenzie, 1954, 1958; Sharpe, 1967; Denver and Hands, 1997; Bochel and Denver, 1971, 1972; Schuman, 1999) this paper brings together context and democratic renewal, attempting to confirm and supplement Pattie and Johnston's (2000) key findings in a place context. This is done by examining individual and grouped voter turnout in a ward through the use of a marked-up electoral roll. The only similar work of this type comes from Taylor (1973) who examined the Victoria ward in Swansea at the 1972 local elections and Dyer and Jordan (1987) in Aberdeen. Utilising returns from one of the party's telling activity (in which party campaigners had asked voters the candidate that they had voted for on leaving the polling station) Taylor was most interested in studying if there was any distance decay effect from where voters lived to polling stations. He found a weak relationship between actual distance and turnout but a much stronger relationship between perceived distance and turnout.

With particular reference to Taylor's and Dyer and Jordan's work, there are a number of ways in which this paper seeks to improve their findings. First, by using the marked-up electoral register used by the polling station administrators we can get a definitive record of who did and did not vote that is not dependent on party campaigners getting voting information from electors (who are not legally obliged to give it to them). Second, a much more accurate framework for analysing network distances, topography and other local aspects can be achieved by constructing a GIS of the ward under investigation. Third, a more complete view of the election can be gained by supplementing the turnout data with data relating to other contextual effects like local party activity and advertising.

The paper presents results from an initial small scale study, examining turnout in a single ward for a single election. The aim of the research is to move towards a more contextual approach to understanding voter turnout behaviour; one that takes into account the different scales at which voting behaviour takes place (the individual, the household, the polling district). The paper uses simplified assumptions regarding spatial voting behaviour as a means of identifying basic trends, although future adaptations are also discussed. Such an approach allows a 'complete' electoral geography to be e one that addresses all reasons for low voter turnout; the social, the administrativ geographical.

17.5 METHODOLOGY

17.5.1 Study Area

The electoral ward of Westbury-on-Trym (or Westbury) in Bristol, UK, was c study area. Westbury was chosen since historically it is one of the few wards that consistently experiences high electoral turnouts. Westbury is sub-divided polling districts referred to as A, B and C. These are shown in Figure 17.1, v shows the location of the polling stations and the road network. The princip district boundary is formed by the A4018, a main thoroughfare linking the Bristol to the M5 to the west. The boundary of A and C is formed by the linking the centre of Westbury with the neighbouring ward of Stoke Bishop. exception of those living in north of polling district A, no voter has to cross a to get to a polling station. Polling district B contains the local shop administrative centre of Westbury. The social and demographic characte Westbury are presented in Table 17.1. This reveals that it has a high percent: and retired people and also people of a high social class. These demogra reflected in the local election results with Westbury being a traditional Co strong hold as presented in Table 17.2.

Figure 17.1 The ward of Westbury-on-Trym in Bristol, UK showing polling districts and polling

Table 17.1 The demographic and socio-economic profile of Westbury (from the 1991 Census of Population).

Males	Females	0-24	25-64	65+
46.31%	53.69%	27.52%	48.33%	24.15%

Occupation	Per cent
Professional	15.64
Managerial	48.56
Skilled Non-manual	15.23
Skilled Manual	9.47
Partly skilled	9.05
Unskilled	0.82
Other	1.23
Retired	59.26

Table 17.2 Local election results for Westbury 1994 – 1998.

Party	1994 %Votes	1995 %Votes	1997 %Votes	1998 %Votes
Conservative	54.84	58.98	53.15	61.68
Labour	31.58	12.60	26.97	18.89
Lib Dem	13.48	12.01	17.24	17.04
Green	-	11.16	2.46	2.27
Spoilt Papers	0.09	5.25	0.18	0.12

17.5.2 GIS and Voting Behaviour

To analyse the geography of voting behaviour at the micro-scale, a GIS was created to map and model local voting patterns. Voting data relating to the 1998 Local Council Elections was acquired for Westbury. The voting data comprised of a marked-up electoral register for Westbury that recorded every person in the ward that had registered to vote, together with information upon whether they had voted in person, voted by postal vote or had not voted at all. These data are kept temporarily on record to avoid electoral fraud and as a means of reference for legal challenges to election results and are in the public domain. Since this data is held at the individual level, it can be used as a means of investigating the effects of location upon voting. This individual data was aggregated into households based upon the addresses of the voters and the percentage of votes per household was then calculated. Postcodes allowed the households to be geo-referenced to Ordnance Survey grid co-ordinates relating to the centroid of the unit postcode. These were then converted into a point coverage which formed the basis of a GIS. Other coverages incorporated into the GIS included road network data, footpath network data, topographical data, and data relating to the boundaries of the three polling districts within the ward. The site of each polling station was also identified.

Socio-economic information relating to the voting population was modelled using Super Profile data. This allocates a socio-economic category to every postcode in the ward. Figure 17.2 shows the geography of Super Profiles in Westbury. The majority of voters are classed as being affluent achievers or thriving greys, reflecting Westbury's status as a wealthy suburb with a high percentage of retired people. The only diversity is to be found to the east of the ward in polling district B, which contains ha families and producers (blue-collar workers).

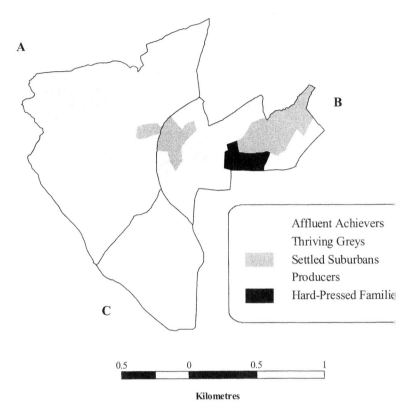

Figure 17.2 The geography of super profiles in Westbury.

Spatial analytical functions in the GIS were used to generate measure geographical context. Distance decay measures were generated using two meth a simple measure of Euclidean distance was generated from each househ corresponding polling station. Second, a more sophisticated measure of dis generated using the topology of the road network and footpaths with the GI calculate the minimum network distance from each household to the polling sta topographical information was used in the GIS to ascertain whether a person ha up or down hill to the polling station. This was estimated by calculating the diff height between each household and the corresponding polling station.

The GIS was used to construct summary statistics regarding the local each polling district. These are shown in Table 17.3. Polling district A is the area although polling district B contains the largest number of registered vote district A also has the voters who, on average, have to travel the farthest. Polli C is by far the smallest and compact with respect to both size and number of vo respect to travel behaviour, polling district A has the least car parking provis polling district B, located in the village centre, has the most opportunities for m destinations.

Table 17.3 Polling district context.

	Area (km^2)	Perimeter (km)	Voters
A	1.90	6.81	3127 (38%)
B	0.92	5.40	3351 (40%)
C	0.71	3.47	1824 (22%)
Westbury	3.53	15.67	8302

	Polling Station	Location	Parking	Multiple Trips	Farthest Point (km)	Average (km)
A	Library	Main road	Average	Poor	1.23	0.67
B	Church	Village centre	Good	Good	1.21	0.48
C	Church	Suburban	Good	Average	0.75	0.34

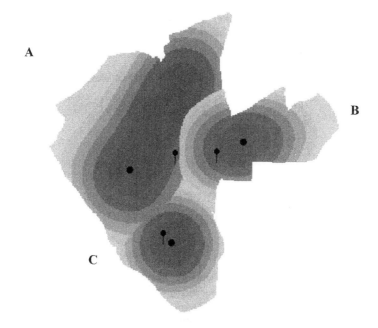

Figure 17.3 A comparison of polling station location with electorate density.

Figure 17.3 shows an electoral density surface of registered voters in each polling district constructed using the GIS. This allows a comparison to be made between the centre of the electoral density in each polling district and the location of each polling station. In terms of optimal location, polling district C has the best-sited station with respect to electoral density whilst A has the worst. It is also important to recognise the influence that the local campaign may have on voting behaviour. Electorates in polling district A had an extra card to remind them to vote a week before the election. This was a minor experiment undertaken by the Electoral Services Unit in the city council to ascertain the affect that an extra reminder has on voting. Finally, the weather on the day of the election was overcast but dry, commonly recognised as encouraging turnout.

17.5.3 Postal Voting

When analysing voting behaviour it is important to distinguish between differen
of voting. Fundamentally, there are three recognised methods: voting in person,
post and voting by proxy in person (when a person registers to vote for someo
their behalf). In the context of this research, there is an important distinctio
postal votes and voting in person since the latter will be more significantly infl
local contextual effects, particularly issues of accessibility to the polling stati
votes tend to be utilised by people who are absent on the day of the election a
with mobility problems (for other reasons see Halfacree and Flowerdew, 1993).

The majority of postal votes within Westbury were for voters living
households (75%), reflecting the bias towards old people living on their own.
postal votes for households of more than two people. At least half the po
represented electorates resident in old people's homes. Table 17.4 shows th
district C has the greatest proportion of postal votes, reflecting the large num
people's homes in this area. Since the marked-up electoral register only indic
person has chosen to vote by post, and not whether they have actually vot
election, this may positively bias the turnout figures if not taken into accou
postal votes were removed from the sample (see also Dyer and Jordan (1987) f
recommendations).

Table 17.4 Voting turnout in the 1998 local elections.

	% Voted	% Postal Votes
A	46	2.13
B	37	4.69
C	47	12.84
Westbury	42.75	5.51

17.6 ANALYSING VOTING BEHVIOUR

17.6.1 Voting by Polling District

Table 17.4 summarises the election turnout for Westbury as a whole and the thr
districts. Overall, 42.75% of the electorate voted compared to an average of
Bristol as a whole indicating that it was a good turnout. When the votes are pro
by polling district, a distinct geography emerges. Whereas polling districts A a
had similar above average turnout, polling district B had significantly less nu
people voting. Broadly this conforms to expectations. Polling district C is the s
terms of area, roughly circular in morphology with the polling station almost co
with the centre of the electorate density and has a resident socio-economic class
high propensity to vote. Hence it is expected that this polling district should
highest turnout. The result for polling district A is a little more surprising
morphology and distribution of voters. However, its favourable socio-economic
respect to the propensity to vote and the fact that it received an extra remi
account for its above average turnout. Polling district B, with its combination
socio-economic classes and elongated morphology has the lowest turnout.

17.6.2 Voting by Household

Pattie and Johnston (2000) have demonstrated that households are the basic unit of analysis with respect to voting for a political party and hence this may also be the case for voting turnout. In particular, we might expect a socialisation effect whereby people within a household encourage or discourage each other to vote. Hence, voting behaviour was analysed by households. Figure 17.4 shows the percentage of votes per one person household in each polling district. It is clear that turnout in one person households is disproportionately higher in C than in other polling districts, reflecting the importance of postal votes of single person households in this polling district. Figure 17.5 shows the affect of removing these votes, with the turnout in polling district C falling in-line with the rest of the polling districts in the ward.

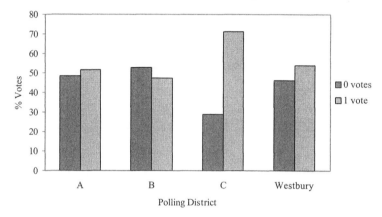

Figure 17.4 Percentage of votes per one person household.

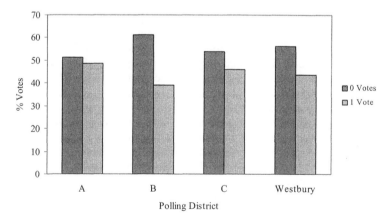

Figure 17.5 Percentage of votes per one person household with postal votes removed.

Table 17.5 shows voting by household size (*i.e.* the number of registered voters resident in the household) by polling district with postal votes removed. This reveals two aspects of voting behaviour at the household level. First, there is a clear decline in voting in all three polling districts as the size of the household gets larger. Second, within households there appears to be a 'dual-voter' effect. This is the phenomenon that occurs in households of two or more people, in which two people in the household have propensity to vote than one person. In other words, people are more likely to vo than vote on their own and so there are less than the expected number of single households of two, three and four voters.

Table 17.5 Percentage of people voting by household size.

		0 Votes	1 Vote	2 Votes	3 Votes
1 person household	A	51.23	48.78		
	B	60.98	39.02		
	C	53.95	46.05		
	Westbury	**56.35**	**43.65**		
2 person household	A	42.64	14.37	42.99	
	B	56.10	13.85	30.06	
	C	45.73	15.70	38.57	
	Westbury	**48.60**	**14.43**	**36.97**	
3 person household	A	42.28	18.12	23.49	16.11
	B	57.69	12.18	19.23	10.90
	C	38.24	18.63	29.41	13.73
	Westbury	**47.17**	**15.97**	**23.34**	**13.51**
4 person household	A	42.42	19.70	25.76	10.61
	B	46.15	15.38	28.85	7.69
	C	48.57	8.57	28.57	14.29
	Westbury	**45.10**	**15.70**	**27.45**	**10.46**

17.6.3 Contextual Effects on Voting Behaviour

To understand the effects of local context upon voting behaviour in more geographical variation in the percentage of votes per household was analysed wi to differences in socio-economic class, accessibility to the polling station and to Table 17.6 is a summary of the parameters of a regression model of percentag per household against socio-economic class. The omitted base variable i achievers. It can be seen that there is no significant difference in the propensi between the omitted variable of affluent achievers and thriving greys, but the p to vote becomes significantly less (at the 5% level) in areas of lower socio-class. Hence the propensity to vote is almost 10% less in areas of settled subu hard-pressed families. Interestingly, settled suburbans have a lower propensit than producers do, but this may be because they live farther from the polling sta reflect a distance decay effect.

Table 17.6 The effect of socio-economic characteristics upon voting.

Predictor	Coefficient	SE	T-Statist
Constant	40.078	1.077	37.22
Thriving Greys	0.031	1.474	0.02
Settled Suburbans	-9.638	3.455	-2.79
Producers	-5.305	2.532	-2.10
Hard-Pressed Families	-9.252	3.158	-2.93

Table 17.7 summarises the parameters of a regression model of percentage of votes per household against Euclidean distance controlling for differences in socio-economic class. The model reveals that, after controlling for distance, settled suburbans propensity to vote has increased by almost 2.5%. In addition, distance has a significant influence on voting behaviour, decreasing the propensity to vote by 5.4% per kilometre.

Table 17.7 The effect of Euclidean distance on voting.

Predictor	Coefficient	SE	T-Statistic
Constant	42.775	1.626	26.31
Thriving Greys	0.131	1.474	0.09
Settled Suburbans	-7.229	3.621	-2.00
Producers	-5.427	2.532	-2.14
Hard-Pressed Families	-9.482	3.158	-3.00
Euclidean	-0.005412	0.002445	-2.21

Table 17.8 summarises the parameters of a regression model of percentage of votes per household against network distance controlling for differences in socio-economic class. The model reveals that network distance has a much stronger effect on voting behaviour (T-statistic is double that of Euclidean distance in Table 17.7) than simple Euclidean distance, reducing the propensity to vote by 8% per kilometre. Moreover, the propensity of settled suburbans to vote becomes similar to that of thriving greys and affluent achievers (T-statistic is not significant at 5% level) implying that the initial reduction in the propensity to vote associated with settled suburbans in the previous model was actually a bias introduced by the use Euclidean distance as an inadequate measure of voting behaviour. Hence, it is only hard-pressed families and producers that vary in their voting behaviour, being less likely to vote than the other residents in the ward.

Table 17.8 The effect of network distance on voting.

Predictor	Coefficient	SE	T-Statistic
Constant	46.416	1.706	27.20
Thriving Greys	-0.416	1.473	-0.28
Settled Suburbans	-6.182	3.522	-1.76
Producers	-5.971	2.530	-2.36
Hard-Pressed Families	-10.459	3.161	-3.31
Network Distances	-0.008057	0.001685	-4.78

Table 17.9 The effect of topography on voting.

Predictor	Coefficient	SE	T-Statistic
Constant	44.657	1.24	36.05
Thriving Greys	-2.130	1.50	-1.42
Settled Suburbans	-11.29	3.44	-3.28
Producers	-10.43	2.61	-3.99
Hard-Pressed Families	-9.56	3.14	-3.04
Topography	-0.368	0.050	-7.38

Table 17.9 summarises the effects of topography on the percentage of household after taking into account the differences caused by socio-economi reveals that topography has a very strong effect upon voting, reducing the t 3.6% for every ten metres difference in height between the polling statio household. This effect is strongest in polling district A, which contains the hig in the ward.

Table 17.10 shows the results of a regression model examining the re between network distance and propensity to vote between the three polling Within each polling district, there tends to be very little variation in the socio-class of voters and hence most of the Super Profile variables are insignific models. In the larger of the two polling districts, network distance has the grea on the propensity to vote, reducing the vote by over 7% per kilometre. This is per kilometre in the smaller and more compact polling district C.

Table 17.10 The effect of network distance on voting by polling district.

Polling District A

Predictor	Coefficient	SE	T-Statisti
Constant	54.27	3.17	17.12
Thriving Greys	-1.01	2.28	-0.44
Producers	-13.84	9.78	-1.42
Network Distance	-0.0079	0.00275	-2.87

Polling District B

Predictor	Coefficient	SE	T-Statisti
Constant	42.13	13.72	3.07
Thriving Greys	-2.54	13.57	-0.19
Settled Suburbans	-2.8	14.01	-0.20
Producers	-2.23	13.7	-0.16
Hard-Pressed Families	-6.65	13.81	-0.48
Network Distance	-0.0073	0.0035	-2.06

Polling District C

Predictor	Coefficient	SE	T-Statist
Constant	58.293	2.465	23.65
Network Distance	-0.0036	0.00324	-11.08

17.7 DISCUSSION

The factors affecting the turnout in local elections are complex. The res indicated that turnout is influence by household size, polling district context geographical factors. Household size is the most important factor influe propensity to vote, with the number of votes decreasing with household size. was that twice as many voters did not turnout in household of four voters than t In addition, the dual-voter effect suggests that households socialise each other not to vote, supporting Pattie and Johnston (2000) research. The result is a significant decline in single voter turnout in households of two voters or more.

Polling district context is important since it has been shown that the smallest and most compact polling district (C) also has the largest turnout. Local campaigning also appears to have an affect, indicated by the comparatively high turnout in polling district A given its morphology. The fact that voters in this polling district had an extra card a few days before the election reminding them to vote may have increased overall turnout in this district.

Perhaps the most interesting results are those associated with the local geography of voting. It has been shown that factors such as social class, distance to the polling station and topography are all influential in the propensity to vote. The effect of social class on voting is well documented and the research confirmed the decline in turnout associated with areas of lower social class. With respect to distance, the research supports Taylor's (1973) findings of a negative relationship between distance to the polling station and voting turnout but also quantifies this reduction. A decrease of 8% per kilometre is a significant reduction in turnout, one that may be crucial in election in a marginal ward. Although on average the distance travelled by a voter was actually quite small (an average of half a kilometre in the ward as a whole), the fact that network distance has a far stronger affect on turnout than Euclidean distance suggests that voters are particularly sensitive in their travel behaviour. This is exemplified by the fact that network distance has the least effect on voting in the most compact polling district (C). The sensitivity to local geography is also highlighted by the significant influence of topography on voting turnout.

17.8 CONCLUSIONS

This research has shown that geography is important in the study of election turnout. Although only a small scale study, it has highlighted some interesting features. The use of a GIS has allowed the interaction between voting behaviour and local context to begin to be unravelled. It would seem that the propensity to vote is partly a function of both household size and local geographical context. Not only is there a dual-voter effect but the distance a voter has to travel and the topography they encounter also influences their voting behaviour. The voter also appears to be very sensitive to local context and even small variations in distance travelled would seem to make a difference. However, the proportion of the variation in turnout explained by these models was relatively small, indicating that more influential factors exist that were not included. These factors are probably related to issues of voter apathy and a general disillusionment with mainstream politics.

The research has been undertaken using simplified assumptions and measurements. The geographical distribution of different sized households has not been taken into consideration when measuring the affect of local geographical context on turnout. Since it has been shown that the propensity for a person to vote is influenced by household size, the distribution of households may become important. Geographical factors such as distance, topography and social class may also interact, potentially confounding their effects on turnout. A possible remedy would be to utilise the GIS to construct an aggregate measure of distance, topography and social class that could also include other factors such as the number of turns that a voter makes when travelling to the polling station. In addition, developing recent work in spatial behaviourism (*e.g.* Golledge and Stimson, 1997) will provide a much more nuanced idea of 'perceived' distances to polling stations than solely using simple network distances. A final limitation is that no account

has been made to contextualise the voter at the individual level. In this respec
age, mobility and other characteristics such as education may become infl
determining propensity to vote. However, such information would be very c
collect without recourse to a much larger survey

To conclude, electoral participation is an issue that is becoming an
concern to both academics and political parties. Voter turnout is generally
particularly amongst certain sections of society (for instance, the young and
voters in particular). There are numerous reasons for this, such as voter apa
general disillusionment with mainstream politics. However, this rese
demonstrated that administrative factors are also influential and these have lo
implications. These include the siting of polling stations, the demarcation <
district boundaries and the importance of local campaigning. Issues of account
'Best Value' will undoubtedly become more important as voter turnout incre:
political significance. Hence it can be argued that methods and techniques, suc
that can be used to measure and monitor voter turnout and facilitate the impleme
policy will become an important part in administrating future elections.

17.9 REFERENCES

Agnew, J.A., 1987, *Place and Politics: The Geographical Mediation of
Society*. (London: Unwin Hyman).
Agnew, J.A., 1996, Mapping politics: how context counts in electoral g
Political Geography, **15**, pp. 129-146.
Bochel, J.M. and Denver, D., 1971, Canvassing, turnout and party su
experiment. *British Journal of Political Science*, **1**, July, pp. 257-269.
Bochel, J.M. and Denver, D., 1972, The impact of campaigning on the result
government elections. *British Journal of Political Science*, **2**, pp. 239-244.
Books, J.W. and Prysby, C.L., 1991, *Political Behavior and the Local Con*
York: Praeger).
Books, J.W. and Prysby, C.L., 1999, Contextual effects on retrospective
evaluation: the impact of the state and local economy. *Political Behavior*,
16.
Cox, K.R., 1969, The voting decision in a spatial context. *Progress i
Geography*, **1**, pp. 81-117.
Denver, D. and Hands, G., 1997, *Modern Constituency Electioneering*. (Londor
Dyer, M.C. and Jordan, A.G., 1987, *Who votes in Aberdeen? Marked Elector
as a Data Source*. Working paper, University of Aberdeen.
Golledge, R.G. and Stimson, R.J., 1997, *Spatial Behavior: A Geographic P*
(London: Guilford).
Halfacree, K. and Flowerdew, F., 1993, The relationship between inter-co
migration and postal voting. *Electoral Studies*, **12**, pp. 247-252.
Holt, R.T. and Turner, J.E., 1968, *Political Parties in Action: The Battle of Bar*
(New York: The Free Press).
Huckfeldt, R. and Sprague, J., 1995, *Citizens, Politics, and Social Comm
Information and Influence in an Election Campaign*. (Cambridge Univer:
New York).
Johnston, R.J., 1991, *A Question of Place: Exploring the practice of Human G
(London: Blackwell).
Johnston, R.J. Pattie, C.J. and Allsopp, J.G., 1988, *A Nation Dividing? The
Map of Great Britain 1979 – 1987*. (London: Longman).

McAllister, I. and Studlar, D., 1992, Region and voting in Britain: territorial polarization or artefact? *American Journal of Political Science*, **36**, pp. 168-199.

Miller, W.L., 1977, *Electoral Dynamics in Britain since 1918*. (London: Macmillan).

Miller, W.L., 1988, *Irrelevant Elections*. (Oxford: Oxford University Press).

Milne, R.S. and Mackenzie, H.C., 1954, *Straight Fight. A study of voting behaviour in the constituency of Bristol North-East at the General Election of 1951*. (London: Hansard Society).

Milne, R.S. and Mackenzie, H.C., 1958, *Marginal Seat. A study of voting behaviour in the constituency of Bristol North-East at the General Election of 1955*. (London: Hansard Society).

Pattie, C.J. and Johnston, R. J., 1995, 'Its not like that round here': region, economic evaluations and voting at the 1992 British General Election. *European Journal of Political Research*, **28**, pp. 1-32.

Pattie, C.J. and Johnston, R.J., 1999, Context, conversation and conviction: social networks and voting at the 1992 British general election. *Political Studies*, **XLVII**, pp. 877-889.

Pattie, C.J. and Johnston, R.J., 2000, "People who talk together vote together": an exploration of contextual effects in Great Britain. *Annals of the Association of American Geographers* , **19**, pp. 41-66.

Pattie, C.J., Dorling, D. and Johnston, R.J., 1997, The electoral politics of recession: local economic conditions, public perceptions, and the economic vote in the 1992 general election. *Transactions of the Institute of British Geographers NS*, **22**, pp. 147-161.

Pattie, C.J., Johnston, R.J. and Fieldhouse, E.A., 1995, Winning the local vote: the effectiveness of constituency campaign spending in Great Britain, 1983–1992. *American Political Science Review*, **89,** pp. 969-983.

Rallings, C. and Thrasher, M., 1990, Turnout in English local elections: an aggregate analysis with electoral and contextual data. *Electoral Studies*, **9**, pp. 79-91.

Rallings, C. and Thrasher, M., 1994, *Explaining Election Turnout*. (London: HMSO).

Rallings, C. and Thrasher, M., 1997, *Local Elections in Britain*. (London: Routledge).

Rallings, C. Temple, M. and Thrasher, M., 1994, *Community Identity and Participation in Local Democracy*. (London: Commission for Local Democracy).

Rallings, C., Thrasher, M. and Downe, J., 1996, *Enhancing Electoral Turnout – A Guide to Current Practice and Future Reform*. (York: Jospeh Rowntree Foundation).

Rose, R. and McAllister, I., 1990, *The Loyalties of Voters*. (London: Sage).

Schuman, A.W.E., 1999, *Boundary Changes, Local Political Activism and the Importance of the Electoral Ward: An Electoral Geography of Bristol 1996 – 1999*. (Bristol: Unpublished PhD thesis, University of Bristol).

Sharpe, L.J., 1967, *Voting in Cities: The 1964 Borough Elections*. (London: Macmillan).

Taylor, A., 1973, Journey time, perceived distance, and electoral turnout - Victoria Ward, Swansea. *Area*, **5**, pp. 59-62.

Taylor, P.J. and Johnston, R.J., 1979, *Geography of Elections*. (Harmondsworth: Penguin).

Thrift, N.J., 1983, On the determination of social action in space and time. *Society and Space*, **1**, pp. 23-57.

Zuckerman, A.S., Kotler-Berkowitz, L.A. and Swaine, L.A., 1998, Anchoring political preferences: the structural bases of stable electoral decisions and political attitudes in Britain. *European Journal of Political Research*, **33**, pp. 285-321.

Zukerman, A.S., Valentino, N.A. and Zuckerman, E.W., 1994, A structural theory of vote choice: social and political networks and electoral flows in Britain and the United States. *Journal of Politics*, **56**, pp. 1008-1033.

18

Towards a European peripherality index

Carsten Schürmann and Ahmed Talaat

18.1 INTRODUCTION

Article 2 of the Maastricht Treaty states as the goals of the European Union, *inter alia,* the promotion of harmonious and balanced economic development, convergence of economic performance, improvement of the quality of life and economic and social coherence between the member states. The proposed Trans-European Transport Networks (TETN) will play a prominent role in achieving these goals as they are to link landlocked and peripheral areas with the central areas of the Community. The identification of those peripheral regions, whose accessibility and transport infrastructure systems are to be improved, is becoming of great political importance. This is underlined by the European Commission's *Cohesion Report* (1997) which emphasises that regions should measure policy success, regularly monitor results and regularly inform the public and political authorities of progress.

This paper presents the results of a *Study on Peripherality* undertaken for DG XVI, Regional Policy, of the European Commission. The purpose was to undertake, for the fifteen EU states and twelve candidate countries, the calculation of an index of peripherality of the 'potential' type. The economic potential of a region is the total of destinations in all regions weighted by a function of distance from the origin region. It is assumed that the potential for economic activity at any location is a function both of its proximity or 'travel time' to other economic centres and of its economic size or 'mass'. The influence of each economic centre on any other centre is assumed to be proportional to its volume of economic activity and inversely proportional to a function of the distance between them. The economic potential of a given location is found by summing the influence on it of all other centres.

The calculation of such peripherality indices involves the acquisition, integration, storage and analysis of various spatial and statistical data sets. The use of GIS functions and techniques was indispensable in conducting such calculations. GIS capabilities and techniques, such as overlaying, network analysis, geo-database management, statistics and presentation were comprehensively utilised demonstrating the benefits of the use of GIS in the field of measuring locations peripherality. For this purpose, an integrated GIS-based *European Peripherality Index* software system was developed to facilitate the calculation of peripherality indices, scenarios comparison, data updating and results demonstration.

The paper presents selected results of the study, compares the different per indices, explains the software system developed and concludes with sugge further refinements of the methodology.

18.2 PERIPHERALITY INDICATORS

Fundamentally, a peripherality indicator can be interpreted as an inverse fu accessibility, *i.e.* the higher the accessibility, the less peripheral a region is lo *vice versa*. Accessibility indicators can be used to analyse peripherality in seve regions can be classified into central and peripheral regions, impacts of differ measures such as transport investments can be evaluated, or impacts of acces: regional development can be analysed. The accessibility indicators used in this based on the assumption that the attraction of a destination increases with declines with distance or travel time or cost. Therefore both size and di destinations are taken into account. The size of the destination is usually repre regional population or some economic indicator such as total regional gross product (GDP) or total regional income. The activity function may be linea linear. Occasionally the attraction term W_j is weighted by an exponent α greater to take account of agglomeration effects, *i.e.* the fact that larger facilities disproportionally more attractive than smaller ones. One example is the attracti large shopping centres which attract more customers than several smaller together match the large centre in size. The impedance function is non-linear. 1 main idea of the so called potential accessibility (Hansen, 1959; Keeble *et* 1988; Schürmann *et al.*, 1997; Schürmann and Talaat, 2000a; Wegener *et a* Generally a negative exponential function is used in which a large parameter β that nearby destinations are given greater weight than remote ones.

The mathematical formula that calculates the accessibility A for a region regions *j* can be expressed as follows:

$$A_i = \sum_j W_j^{\alpha} \exp(-\beta c_{ij})$$

where c_{ij} is the generalised cost of reaching region *j* from region *i*.

Potential indicators are frequently expressed in percent of average acces: all regions or, if changes of accessibility are studied, in percent of average acces: all regions in the base year of the comparison.

The model developed is capable of calculating a large number of differe indicators. The range of indicators available will be briefly explained in the paragraphs.

18.2.1 Spatial Aggregation

All calculations of peripherality indices are based on level 3 of the Nomen Territorial Units for Statistics (NUTS) defined by Eurostat and are then agg levels 2, 1 and 0 of the NUTS for the EU member states (Eurostat, 1999a) and geographical units as identified by Eurostat for the candidate and EFTA

(Eurostat, 1999b) by averaging over NUTS-3 regions weighted by NUTS-3 region population. The smallest unit available for this study, *i.e.* the NUTS 3 level, represents counties or local authority regions, whereas NUTS 0 level represents countries; the two levels in between these extremes represent other or standard regions, depending on the country considered.

18.2.2 Modes

Since speed limits for cars and trucks differ and statutory drivers' resting periods affect freight transport, all indicators were calculated separately for passenger and freight road transport. Travel time matrices and peripherality indices for cars represent the perspective of service firms and consumers, namely how many opportunities, such as clients, markets or tourist facilities can be reached from location. Travel time matrices and peripherality indicators for lorries, *i.e.* for goods transport, can be interpreted from the perspective of producers on (potential) markets as the answer to the question which location has the highest market potential.

18.2.3 Mass Terms

Peripherality indices are calculated for each origin region by adding up the mass of each destination region weighted by a function of distance from the origin region. Usually, the mass is measured in terms of gross domestic product (GDP). In this study, also GDP in purchasing power standards (PPS), employment and population are used as mass terms. Distance is measured as the average travel time from one region to every other region in the form of a matrix. The regions are represented by their 'centroids', *i.e.* their main urban centres. Statistical data representing the four mass terms were compiled from Eurostat (1997) and linked to the relevant regions in an integrated database.

18.2.4 Type of Indicator

All peripherality indices are derivatives of potential accessibility. Two different types of peripherality indices are defined:

- Peripherality Index 1 (PI1): The region with the highest potential accessibility, *i.e.* the most central region, is defined to have a peripherality index of zero. The region with the lowest potential accessibility, *i.e.* the most remote region, is defined to have a peripherality index of one hundred. The peripherality index of all other regions is a linear interpolation between zero and one hundred proportional to their potential accessibility. The higher the peripherality index, the higher the peripherality.

- Peripherality Index 2 (PI2): The average potential accessibility of all regions weighted by regional population is defined to be one hundred. The peripherality index of all regions is calculated as potential accessibility expressed in percent of average accessibility. The higher the peripherality index, the lower the peripherality. Peripherality Index 2 is therefore in fact a standardised accessibility indicator.

18.2.5 Spatial Scope of Standardisation

The standardisation was done for three different territories: EU member s
member states plus five candidate countries (Estonia, Poland, Czech Republic,
Slovenia) and EU member states plus twelve candidate countries (the five
above plus Latvia, Lithuania, Slovakia, Romania, Bulgaria, Cyprus, Malta). Th
that the values of the regional peripherality indices differ depending on the
covered.

Based on the above classification (4 NUTS levels, 2 modes, 4 mass term
of indicators, 3 territories), 4 x 2 x 4 x 2 x 3 = 192 possible output indicatc
calculated and mapped. Table 18.1 summarises the 192 possible output indica
numbers in the table are consecutive numbers used for identifying the resulting
output files (see Schürmann and Talaat, 2000b).

Table 18.1 Available output.

Mode	NUTS level	Standardisation	EU member states				EU plus 5 candidates				EU plus 12 can		
			Population	Employment	GDP	GDP in PPS	Population	Employment	GDP	GDP in PPS	Population	Employ-ment	
Car	0	PI1	1	2	3	4	9	10	11	12	17	18	
		PI2	5	6	7	8	13	14	15	16	21	22	
	1	PI1	25	26	27	28	33	34	35	36	41	42	
		PI2	29	30	31	32	37	38	39	40	45	46	
	2	PI1	49	50	51	52	57	58	59	60	65	66	
		PI2	53	54	55	56	61	62	63	64	69	70	
	3	PI1	73	74	75	76	81	82	83	84	89	90	
		PI2	77	78	79	80	85	86	87	88	93	94	
Lorry	0	PI1	97	98	99	100	105	106	107	108	113	114	▮
		PI2	101	102	103	104	109	110	111	112	117	118	▮
	1	PI1	121	122	123	124	129	130	131	132	137	138	▮
		PI2	125	126	127	128	133	134	135	136	141	142	▮
	2	PI1	145	146	147	148	153	154	155	156	161	162	▮
		PI2	149	150	151	152	157	158	159	160	165	166	▮
	3	PI1	169	170	171	172	177	178	179	180	185	186	▮
		PI2	173	174	175	176	181	182	183	184	189	190	▮

18.3 IMPLEMENTATION IN ARCINFO

A software system was developed based on ESRI's ArcInfo to calculate the
Peripherality Index (E.P.I.) which consists of several macros written in A
Language (AML) (Schürmann and Talaat, 2000b). The E.P.I. is intended to fac
updating of input data, the definition of scenarios, the calculation of peripherali
and the presentation of results.

The system itself consists of three core components: the INITIAL macro, t
macro and the PLOT macro. The INITIAL macro defines global variables and p
and initialises ASCII input files. The CALCUL macro calculates the indicators pr

Table 18.1, stores them in output coverages and INFO tables and exports them into ASCII files. The PLOT macro is used to produce output maps showing the resulting peripherality indices. In addition, a number of add-ons were developed to support certain tasks for updating the geodatabase or to perform error checking. As a principle, updates of the database or changes of the parameters can be achieved by editing the input coverages and parameter files. It is not necessary to edit the code of the macros itself to change default settings. The User Manual (Schürmann and Talaat, 2000b) gives a comprehensive description of the model system, explains how to run the system and how to edit the database and suggests how to handle errors.

The CALCUL macro is the central core macro for the calculation of the peripherality indices. It calculates all values for all indicators, modes, levels, territories and masses in one run. Such calculation is conducted within the CALCUL macro according to the following steps:

1. Link travel times for all links in the input network coverage are calculated or updated based on national speed limits as initialised by the INITIAL macro. The basic road network coverage used in this study was extracted from the European road network database developed by IRPUD (1999). The national speed limits for cars and lorries for urban roads, major roads, expressways and motorways were compiled from ADAC (2000), IRU (2000) and UBA (1998). Moreover, link travel times take account of speed constraints in urban and mountainous areas and sea journeys. Speed limits and congestion in urban areas are estimated as a function of population density at NUTS-3 level. It is assumed that the higher the population density, the slower will be the speeds. Road gradients are estimated by overlaying the road network with a digital terrain model (DTM) extracted from the U.S. Geological Survey (2000).

2. As a pre-requisite of calculating accessibility indicators, travel time matrices are calculated based on average travel times between all NUTS-3 regions. These travel time matrices are used to calculate regional accessibility indicators, which are then converted to peripherality indices. Shortest path analysis models offered by ArcInfo were utilised to derive such travel time matrices. Beyond link travel times, the travel time matrices take account of border delays and ferry boarding times, and, in case of freight transport, statutory drivers' resting periods. The matrices are derived by applying the nodedistance command between all regions' centroids and with the link travel times as link impedances and border waiting times and ferry boarding times as node impedances.

3. Based on the travel time matrices, accessibility indicators are calculated for NUTS-3 regions using formula 18.1.

4. Peripherality indices for the three different spatial scopes for standardisation are then calculated and standardised for the NUTS-3 level.

5. The travel times between NUTS-3 regions are then aggregated to travel times between NUTS-2, 1 and 0 regions as weighted averages over NUTS-3 regions.

6. Statistical spatial analyses offered by ArcInfo were run over the obtained accessibility indicators for NUTS-3 region in order to aggregate values to levels 2, 1 and 0 of the NUTS by averaging over NUTS-3 regions weighted by population.

7. The peripherality indices for all NUTS-0, 1 and 2 regions considered are then calculated and standardised.

18.4 THE E.P.I. SYSTEM

The first two macros of the core components of the E.P.I software, *i.e.* the IN
CALCUL macros, run automatically, whereas the PLOT macro depends on contin
interaction. The INITIAL macro does not require user input after the macro is s
CALCUL macro invokes a selection menu after its start, but after all options ;
further user input is requested; in case of the PLOT macro, a windows-based
menu is permanently invoked until the macro is stopped.

As the first step to calculate the required indicators, the INITIAL macro is
The macro sets global variables, parameters and path names necessary t
calculations, reads and initialises ASCII input parameter files and establishe
INFO tables.

As a second step, the CALCUL macro is run to calculate all 192 peripheralit
After its execution a selection menu is called at which the input coverages a
options are selected (Figure 18.1). Two input coverages must be specified,
network and the regions coverage. Then, the reference year has to be chosen.
can choose to assess the peripherality at the base year 2000 or to apply a futur
for 2016 (which is the target year for full implementation of the TEN progr
third selection refers to the output options, which determine the way the res
E.P.I. calculations are stored and presented to the user. The standard output o
present the indices as ArcInfo coverage. The coverage will always be generate
option cannot be unchecked. Additionally, there are three output options availa
can be checked or unchecked in any combination. These are: (i) *Travel time*
(ASCII files) so that travel time matrices are exported to ASCII files, (ii) *Tr*
matrices (INFO tables) so that travel time matrices are maintained as INFO tabl
Peripherality indices (ASCII files), so that peripherality indices are exported
files.

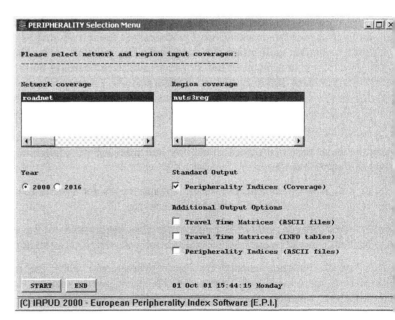

Figure 18.1 Menu interface of the **CALCUL** macro of the E.P.I. software.

These options enable or disable output of the travel time matrices and peripherality indices depending on which results are of interest and how they are going to be further processed. The standard output coverage enables full ArcInfo capabilities with respect to further analysis or way of presentation. However, if the results are to be integrated into a document or into further analysis using other software, the internal ArcInfo formats might not be appropriate. In that case, ASCII files may be generated which enable data exchange between different hardware platforms and software systems.

When the index calculations finish, a report is displayed (Figure 18.2) which summarises input and output coverages as well as the output ASCII files created.

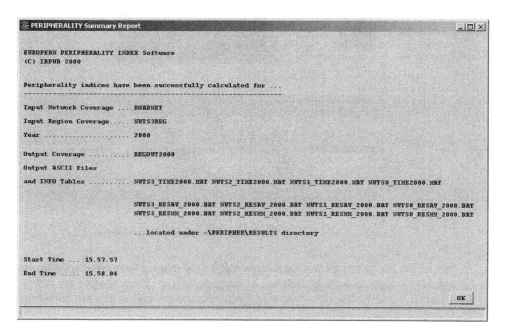

Figure 18.2 CALCUL macro summary report menu.

Finally, the PLOT macro is run to allow the display and presentation of the final results of peripherality indices produced by the CALCUL macro. The PLOT macro invokes an easy-to-use user interface (Figure 18.3) which enables the user to set the combinations of peripherality parameters to be displayed (map parameters) or to select a map by its number according to the naming convention described earlier in Table 18.1. Map parameters are the parameters used to calculate peripherality indices. There are five parameters which determine peripherality index calculations: *Index, Mode, Level, Territory* and *Mass*. For each parameter several options exist. Each combination of options produces a different peripherality index map. In total, it is possible to produce 192 different maps based on 192 parameter combinations. After the map parameters are set, the composed map can be displayed, printed and/or exported to a different format. Available export format options are *png, tiff* or *ai*. The user can also choose between A4 and A0 output size.

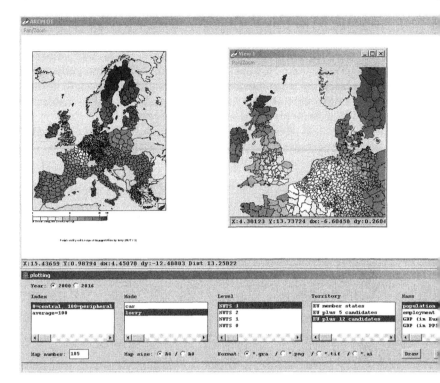

Figure 18.3 User-interface of the plotting macro of the E.P.I. software.

Apart from the above three core macros, additional macros were develop
E.P.I. software system to support updating of the geodatabase. These macros a
if the road network is edited, new centroids are chosen or new region data are i
to the region coverage.

18.5 RESULTS

The results showed that the general spatial patterns of peripherality are ver
across all indicators calculated, reflecting the fact that distant geographica
cannot be fully compensated by transport infrastructure (Schürmann and Talaa
However, each indicator emphasises certain aspects of peripherality. So, the ch
type of peripherality index to be used becomes a matter of concern. Dependi
purpose of the study, a certain indicator type may be more appropriate than an
as certain subsets of regions yield slightly better or worse results with r
peripherality. As an example, Plate 8 shows the peripherality index with respec
by lorry for NUTS 3 level regions. Regions in the European core show th
accessibility and so are most central. These regions are located along the 'Blu
in western Germany, Belgium, in the southern parts of the Netherlands, in
France and in southern England. A band of regions from the Po estuary towa
and Lyon up to the Channel coast show above-average accessibility. Regions in
countries, in Romania and Bulgaria show index values of less than 10, *i.e.*

peripheral regions, due to their – compared to EU member states – still relatively poor economic performance.

Based on other studies (Fürst *et al.*, 2000; Schürmann *et al.*, 1997) and on theoretical considerations, the peripherality index shown in Plate 8 together with peripherality with respect to population by car were proposed as the standard peripherality indices. The correlation of these two indicators confirms a high degree of similarity (Plate 7). Although the overall correlation seems clear, there are some small, but nevertheless important differences between both indices. In general, central regions in Benelux, Germany and France, but also in the UK show comparably higher values for peripherality with respect to GDP than with respect to population. On the other side, regions in the candidate countries have higher accessibility to population than to GDP. This is because most of the candidate countries have relatively poor economic performance but large populations which confirms the observation that peripherality index with respect to population seems less polarised than peripherality with respect to GDP. In other words, if peripherality with respect to GDP is used, central regions appear less peripheral; if peripherality with respect to population is used the candidate countries appear less peripheral.

However, a number of comparisons between the other peripherality indices showed that the choice of indicator has great influence on the results. The overall spatial patterns of all peripherality indices are very similar, so correlation between different indicators are rather high. This reflects the fact that, irrespective of the kind of peripherality index used, the distant geographical position of peripheral regions cannot be fully removed by transport infrastructure improvements.

Peripherality with respect to population by car is less polarised than peripherality with respect to GDP by lorry reflecting the fact that population is more evenly distributed across Europe and that because of faster driving speeds for cars a greater number of opportunities can be reached. On the other hand, peripherality with respect to lorry favours regions around the Channel coast, since for lorries (*i.e.* freight transport) the 'barrier effect' of the Channel is much less than for cars (*i.e.* passenger movements), because for private trips the Channel seems to be a greater obstacle. Candidate countries benefit more if peripherality with respect to population by car is used; conversely, central regions benefit more if peripherality with respect to GDP by lorry is used.

The type of indicator has relatively little influence on the results. Standardisation between the minimum and maximum shows slightly more differentiation among peripheral regions, whereas standardisation on the European average shows slightly more polarisation between the central regions. The greater the territory used for standardisation is (*i.e.* the more candidate countries are taken into account), the lower is the European average, and the more will regions in EU member states improve their relative position.

Due to the overall objective of using purchasing power standards, GDP in PPS has slight balancing effects compared to GDP in Euro; nevertheless, peripherality with respect to both is more polarised than peripherality with respect to population or employment.

Finally, the higher the NUTS level, the greater is the loss in spatial differentiation. Studies based on the NUTS-3 level yield a great number of detail and differentiation between and within peripheral and central regions. This is particularly true for the relatively small German, French and Italian regions.

18.6 ROLE OF GIS

With its functionality, the software system can be seen as a GIS-based contrib
wider system for measuring and monitoring the success of EU policies with
peripherality and cohesion. The use of GIS has enhanced the accessibility
process in various ways. The integration between spatial and statistical
different sources in one unified database allowed easier data accessibility,
maintenance and analysis. The application of the spatial statistics function
resulted in more accurate and faster calculation and aggregation of the ac
indicators. The execution of network analysis models offered by ArcInfo made
to calculate travel times among a very large number of centres. The pr
capabilities of GIS allowed a clear and easy visualisation and interpretati
obtained results. Finally, the development of such integrated system allow
construction of different scenarios and an easy comparison of scenarios.

For calculating distance measures, *i.e.* average road travel times of passe
goods, the developed software takes account of road types, speed limits for
lorries, congestion in urban regions and delays due to mountainous areas
borders and maximum driving hours of lorry drivers. In this the system goes b
way usually travel times are measured in accessibility studies. Moreover, per
indices are calculated for NUTS-0, NUTS-1, NUTS-2 and NUTS-3 regions b
unified and disaggregate approach. That was achieved by incorporating GI:
capabilities along with network analysis functions such as `nodedistance`, `impe`
`turntable`.

Additional strengths of the system can be seen in its flexible modular
which is expandable, in the core macro that calculates all peripherality indic
model run, in the variety of output options available, in the minimised numbe
interactions, in the possibility to run the system under UNIX or Windows NT /
2000, in the fact that all input coverages and input files can be manually edited
or exchanged, in the combination of windows-based menu operations designed
friendliness and command line executions designed for efficiency, and in the ca
provided for designing future scenarios.

The integration of all types of data storage, analysis, updating, present
scenario building in one integrated system, without the need to export/im
to/from another software, is another very important advantage of the use of GI
application.

Compared to these strengths, the software has only little weaknesses. C
relatively long processing time of the core macro which is due to the fact that it
all 192 indicators in one run. Also the relatively large amount of disc space re
temporary coverages and for storing results might limit the applicability of the
the current version, the model considers only road traffic and neglects rail, air a
waterways and so is not able to calculate intermodal accessibilities. Moreo
accessibility of the potential type can be calculated, whereas daily access
average travel costs are not taken into account.

18.7 CONCLUSIONS

The present study showed how the developed system utilises GIS capabilities
and support the assessment of EU policies with respect to peripherality and col

summary, for all kind of indicators, regions in western Germany, northern France, Belgium, the Netherlands, southern England and northern Italy show the highest accessibilities and can be considered the most central regions. When NUTS-3 regions are considered, great differences in peripherality can be found between peripheral regions, for example in Scandinavia, Greece and on the Iberian Peninsula. This indicates that the model is able to capture relatively small, but nevertheless important differences. When higher NUTS levels are considered, these details partly disappear.

The system evaluated the peripherality of EU member states and candidate countries by a number of different peripherality indices with respect to NUTS levels, modes, mass terms, types of indicator and spatial scope of standardisation. The software system developed offers the full range of combinations of these parameters, totalling 192 indicators. The general spatial patterns of peripherality are very similar across all these indicators, reflecting the fact that distant geographical location cannot be fully compensated by transport infrastructure. Each indicator emphasises certain aspects of peripherality. So, the choice of the type of peripherality index to be used becomes a matter of concern. Depending on the purpose of the study, a certain indicator type may be more appropriate than another type.

From a theoretical point of view, it would be of great interest to incorporate also the other modes, namely rail, air and inland waterways, into the system to enable calculations of intermodal accessibilities and peripherality indices. Also of interest would be the possibility to calculate daily accessibility or average travel costs. A more practical extension would be to incorporate a 'scenario manager' which would allow generation, management and application of different scenarios.

18.8 ACKNOWLEDGEMENTS

The theoretical concepts of accessibility and peripherality is based on previous work performed at IRPUD in the EU project *Socio-economic and Spatial Impacts of Transport Infrastructure Investments and Transport System Improvements* (SASI) commissioned by DG VII (Transport) of the European Commission as part of the 4th RTD Framework Programme and in the Working Group 'Geographical Position' of the Study Programme for European Spatial Planning (SPESP) organised by DG XVI.

The definition of the centroids used for the accessibility calculations and the compilation of the socio-economic database used in this study were contributed by Andrew Copus, Rural Policy Group, Scottish Agricultural College (SAC), Aberdeen.

18.9 REFERENCES

Allgemeiner Deutscher Automobil Club (ADAC), 2000, *Tempolimits.* http://www.adac.de.

European Commission, 1997, *Cohesion and the Development Challenge Facing the Lagging Regions.* Fourth Cohesion Report. Regional Development Studies. (Luxembourg: Office for Official Publications of the European Communities).

Eurostat, 1997, *New Chronos Database. Tables d3pop, xe_gdp, pvd0e.* (Luxembourg: Office for Official Publications of the European Communities).

Eurostat, 1999a, *Regions. Nomenclature of Territorial Units for Statistics* (Luxembourg: Office for Official Publications of the European Communities)

Eurostat, 1999b, *Statistical Regions in the EFTA Countries and the Central Countries (CEC).* (Luxembourg: Office for Official Publications of the Communities).

Fürst, F., Schürmann, C., Spiekermann, K. and Wegener, M., 2000, *The SA Demonstration Examples. SASI Deliverable D15.* (Dortmund: Institute Planning).

Hansen, W.G., 1959, How accessibility shapes land-use. *Journal of the American Planners* **25**, pp. 73-76.

International Road Union (IRU) (editor), 2000, *Speed Limits, Goods Trans,* Information Center. http://www.iru.org/IRU_NEWS/IRU/ECOMMEF Ecommerce.E.asp, October 2000.

IRPUD, 1999, *European Transport Networks.* Dortmund: Institute for Spatial http://irpud.raumplanung.uni-dortmund.de/irpud/pro/ten/ten_e.htm

Keeble, D., Offord, J. and Walker, S., 1988, *Peripheral Regions in a Community Member States.* Luxembourg: Office for Official Publications of the Communities.

Keeble, D., Owens, P.L. and Thompson, C., 1982, Regional accessibility and potential in the European Community. *Regional Studies* **16**, pp. 419-432.

Schürmann, C. and Talaat, A., 2000a, *Towards a European Peripherality In report.* Report for General Directorate XVI Regional Policy of the Commission. (Dortmund: Institute of Spatial Planning).

Schürmann, C. and Talaat, A., 2000b, *Towards a European peripherality In Manual.* Report for General Directorate XVI Regional Policy of the Commission. (Dortmund: Institute of Spatial Planning).

Schürmann, C., Spiekermann, K. and Wegener, M., 1997, *Accessibility Indicato and Report.* SASI Deliverable D5. (Dortmund: Institute of Spatial Planning).

Umweltbundesamt (UBA), 1998, *Geschwindigeitsbegrenzungen.* ht umweltbundesamt.de/uba-info-daten/daten/geschwindigkeitsbegrenzung Last updated May 09 1998, (Berlin: UBA).

U.S. Geological Survey, 2000, *GTOTO30 Digital Terrain Model.* EROS Da http://edcdaac.usgs.gov//gtopo30/gtopo30.html.

Wegener, M., Eskelinen, H., Fürst, F., Schürmann, C. and Spiekermann, *Indicators of Geographical Position.* Final Report Part 1 of the Worki 'Geographical Position' of the Study Programme on European Spatial (ESPON). (Bonn: Bundesamt für Bauwesen und Raumordnung). Dra availabe at: http://www.nordregio.se/spespn/Files/ 1.1.final1.pdf.

19

Using a mixed-method approach to investigate the use of GIS within the UK National Health Service

Darren P. Smith, Gary Higgs and Myles I. Gould

19.1 INTRODUCTION

This chapter describes an ESRC-funded research project which is examining the current and potential uses of Geographical Information Systems (GIS) within the UK National Health Service (NHS). We contend that previous questionnaire-based studies in this area have provided partial understandings of this issue but have not, to date, fully teased out the underlying factors which facilitate and/or constrain GIS uptake and use within primary and secondary care. In this paper, we outline the research design that is being used to address this caveat which, in turn, serves to illustrate Gatrell's (2001) call for the adoption of mixed-method research frameworks within health-based research.

In this chapter we describe a mixed-methods research project which is examining the types of factors influencing variations in the use of GIS within the UK NHS. Previous studies, described below, have provided important insights into the current levels of take-up of GIS, but have tended to involve quantitative analysis usually based principally on postal and telephone questionnaire approaches. Our current research project has also used questionnaire surveys, but merely as a context for exploring factors influencing both levels of adoption and current utilisation of GIS, via qualitative techniques based around semi-structured interviews. In addition, we are considering the contribution GIS can make to 'new' policy agendas in the UK, particularly the objectives set out in *The NHS Plan* (Department of Health, 1998) and other national policy documents.

The main aim of this chapter is to discuss the benefits of employing a 'sequential' mixed-method approach in order to investigate the take-up, and application, of GIS in the UK NHS. We define mixed-methods, in line with Philip (1998: 264), as a research process which employs two or more methods "to address a research question at the same stage in the research process, in the same place, and with the same research subjects". In the context of our specific research concerns, we assert that a mixed-method approach will provide a more comprehensive analysis of the use of GIS within the NHS, in order to

fully tease out the interplay between organisational, technical and project ma
barriers that underpin the (non)uptake of GIS in different settings within the NH

The chapter begins by discussing arguments for using mixed-methods a
(Section 2). It then reviews current knowledge and previous studies of the u
within the NHS (Sections 3 and 4). We then discuss our project which is
underway (Section 5) before outlining the mixed-methods research design th
using (Section 6).

19.2 ARGUMENTS FOR A MIXED-METHOD APPROACH

In a recent review of the methods and techniques employed by researchers
geographies, Gatrell (2001) points to a chasm between work which adopts 'mea
led' and 'interpretation-based' approaches. Noting the key differences, Gatrell
the former approach under the broad heading of 'mapping', and this is equ
quantitative spatial data analysis. For purposes of convenience, Gatrell claim
strand of health geography involves a three-fold distinction between the
'visualisation' (identifying spatial patterns), 'exploratory spatial data
(investigation of association between variables to explain spatial patte
'modelling' (formal statistical procedure to test hypothesis). The interpretat
approach, Gatrell notes, involves a range of clearly defined methods to collect c
data and which usually seeks to "understand human beliefs, values and
incorporating the use of interviews, participant observation and focus groups.

These different approaches are illuminated further in a selective su
previous health studies, and this serves to show that: "those researching the geo
health have tended to use one or the other set of tools" (Gatrell, 2001: 82). Im
Gatrell concludes the discussion by problematising the dichotomous relationshi
the quantitative (mapping) and qualitative schools of thought, proclaiming:

> "The danger, however, is that the analytical methods give very little attention to
> the points or dots on the map really represent. The dots are not inanimate objects;
> are real people, and while the ways in which they are arranged spatially may
> some light on disease causation the [quantitative] approaches… give no consider
> at all to the feelings, experiences, beliefs and attitudes that the individuals have.
> need to have these people speak to us rather than reducing them to a collection of
> on a map" (Gatrell, 2001: 78).

To overcome this dualism, and contrary to the above orthodoxy, Gatrell ac
'triangulation' approach; whereby both quantitative and qualitative met
techniques are mixed, and findings integrated to corroborate the findings of o
(see Brannen, 1992 for discussion of the merits of triangulation). To exer
benefits of such a research design, Gatrell draws attention to a number of con
studies:

> "which have taken a more pragmatic and eclectic stance, using whatever method
> appropriate to the problem under investigation" (Gatrell, 2001: 82).

This mixing of multiple methods, Gatrell suggests, has proved valua
insights from in-depth interviews adding colour and explanatory power to q
studies" (*ibid*).

This standpoint parallels similar calls for multi-method research in other areas of human geography. For example, McKendrick (1999), Sporton (1999), Findlay and Li (1999) and Graham (1999) provide insightful discussions of the potential of multi-methods research in population geography.

19.3 CURRENT UNDERSTANDINGS OF THE USE OF GIS WITHIN THE NHS

Overall, the take-up of GIS within the NHS is unclear with disagreements about the levels and types of GIS use within the NHS. Some commentators have demonstrated that there are good examples of GIS uptake within the NHS (*e.g.* Taylor and Allgar, 1997), and claim that many of the 'obstacles of the past' have been removed. For example, emphasis is often made to the reduced financial and staff resources required to implement and maintain GIS, particularly the lower 'start-up' costs and less technical problems associated with GIS, when compared to the 1980s and early 1990s (Smith, 1999). This, coupled with improved access to cheaper, higher quality, digitally coded health data and a growing organisational recognition within the NHS of the need for spatial analysis within health service planning and monitoring, would appear to provide a conducive context for GIS to flourish within the NHS. This point, to a certain extent, is reflected in recent conferences held in the UK and further afield, such as the USA where there is a longer tradition of GIS in the health sector.

By contrast, some authors are less optimistic, and claim that the current implementation and use of GIS within NHS organisations "is still very much in its infancy in the healthcare sector" (Burns, 1996: 37). It is argued that previous forecasts of the widespread uptake of GIS within the NHS (*e.g.* Wrigley, 1991) have not been realised (Barlow, 2000). This contention would appear to have some validity. For example, in a recent review of a leading professional health journal, Smith *et al.* (2001) identified only one example of health-related research which had utilised GIS. Similarly, Higgs and Gould's (2000) analysis of recent NHS policy documents found no mention of the potential for GIS within the 'New NHS'. Previous surveys of GIS use within the NHS also support this notion of low levels of GIS uptake within NHS organisations (Cummins and Rathwell, 1991; Gould, 1992; Smith and Jarvis, 1998; Cooper, 2000). Evidence from other European countries also suggests a 'patchy' and 'uncoordinated' level of GIS adoption in health organisations (*e.g.* Ireland – Houghton, 2001).

But why is there a limited uptake of GIS within the NHS? What are the factors impeding the current and potential use of GIS? How valid are optimistic forecasts of extensive GIS use within the NHS, such as Burns (1996: 39) view that "mapping applications will soon form part of the core suite of office applications, sitting alongside word-processor and spreadsheet packages, and will become an integral part of any manager's desktop".

To comprehensively address current and future uses of GIS, we operationalise (using mixed-methods) the call for enhanced knowledge and understandings of the reasons why the NHS has not fully realised the benefits of GIS-uptake in relation to new technical and organisational opportunities (Higgs and Gould, 2001). The following section now reviews earlier surveys of the use of GIS within the NHS to elucidate some of the factors that are perceived to have hindered GIS uptake within the NHS in the past.

19.4 PREVIOUS STUDIES OF THE USE OF GIS WITHIN THE NHS

An early investigation of the use of GIS within the NHS was commission
Association for Geographic Information (AGI), and carried out by Cumm
Rathwell (1991). This study, using a telephone survey of the 14 Region
Authorities (RHAs) and 190 District Health Authorities (DHAs) in Engla
significant geographical variations of GIS use between, and within, Health
Despite noting a 'surge' in interest in GIS within the health sector during the lat
number of concerns regarding the future uptake of GIS were highlighted. Fi
claimed that GIS uptake was being hindered by a low awareness of the value
spatially representing population and health needs-based information (itself a
requirement of DHAs in 1991). This was attributed to previous low levels of s
handling by health professionals, managers and IT staff within the NHS. Ind
cultural and organisational criteria were discussed, which were seen to partly e
under-utilisation of GIS. In particular, it was noted that health planners
previously allocated health-based resources on the needs of the local popula
therefore geographic information was not seen as an important factor duri
making and health planning activities. In addition, Cummins and Rathwell (19
to a lack of infrastructure to enable staff, training, resource management and
budgets to implement GIS, and assert that the small size of many IT departu
high staff turnover hindered the effective implementation of GIS. This wa
influenced by the lack of understanding of GIS suppliers for the needs
authorities, and how organisations within the NHS could make best use of GIS
Cummins and Rathwell (1991: 14) conclude: "The potential in the
geographically derived information is enormous; it is unfortunate that so far ma
service have yet to realise or appreciate this".

A wider-geographic survey of directors of public health and informatio
officers, in all of the 197 District Health Authorities in England and Wales in
undertaken by Gould (1992). In contrast to Cummins and Rathwell (1991), Go
to a generally high level of awareness of the value of GIS amongst health off
found that their use was largely confined to relatively low-level operational task
desktop mapping, in order to identify local population health needs. GIS was
used for strategic tasks within the NHS, and Gould identified a number
influencing such trends. As a result, two key recommendations are put f
encourage the greater use of GIS within the NHS. First, Gould (1992: 399) str
"something must be done to improve the GIS skills and understanding in the N
use of such systems is to become more widespread". Second, it is argued that "
need to widen the functional capabilities of GIS software for spatial analysis a
modelling". Given the organisational and technical changes that have taken p
Gould's survey, an important aim of the current project has been to update the
findings.

Similarly, when examining the impact of NHS reforms from the early 199
use of GIS, Smith and Jarvis (1998) found relatively low-level, uncoordin
sporadic uses of GIS within the NHS. It was argued that this was partly due to
problems, such as data quality, but that the lack of policies and guidance rega
use of GIS technology across the NHS was also a major contributory factor. M
case study material appears in so-called 'grey literature' and the authors co
espousing the benefits of improved data availability, detailed meta-data, ap
organisational mechanisms, and greater levels of awareness of the potential f

order to promote the exchange of data between agencies (see also, Martin, 1996). Smith and Jarvis also suggest that GIS uptake was being hindered by a limited transfer of ideas between the GIS research community in the academic sector and those based in health departments despite the existence of more informal networks where advice and ideas were being exchanged with external agencies and organisations (see Reeve and Petch, 1999) for a comprehensive discussion of technical, human and organisational factors influencing the effectiveness of GIS).

More recently, a questionnaire-based study of GIS use within the 13 Health Authorities of the West Midlands was undertaken by Cooper (2000). In line with earlier studies, a series of obstacles were identified which were hindering the use of GIS technology across the region. It was noted that two-thirds of the Health Authorities in the West Midlands stated that GIS was being under-utilised and not being fully exploited. The main hurdle was the high costs of 'unaffordable' digital geographical data, which restricted the potential uses of GIS. Other constraints included the lack of resources for training and available work time for NHS personnel, due to the low priority (as deemed by senior management) given to GIS within the organisations. To overcome these problems, Cooper (2000: 33) calls for the establishment of: "Regional health GIS centres, integrated with the newly formed public health laboratories", which "would be able to take advantage of regionally collected health data, achieve economies of scale with regards GI, reduce local duplication of effort and provide GIS support for the local health services". This is an important recommendation that requires further investigation, and which will be fully examined in the course of our research project.

Without doubt, the above studies have provided valuable insights into understanding the current, and projected, roles of GIS within the NHS. Significantly these studies have pointed to sporadic levels of GIS uptake within primary and secondary health care organisations, and have been relatively consistent in identifying the types of factors that are hindering the implementation and use of GIS within the NHS. Despite this contribution, we would contend that more in-depth research is required in order to examine the interplay between constraining and enabling factors in specific contexts and their influence on the level and type of GIS uptake, particularly in light of recent structural changes and new technical developments.

Questionnaire-based studies have arguably failed to fully tease out the processes that underpin the use of GIS within the NHS, a fact that is generally acknowledged by the authors of these studies who, in turn, have called for more in-depth studies to explore such issues. In order to address these shortcomings, we would concur with Philip (1998: 273), who suggests that:

> "researchers should think beyond the myopic quantitative – qualitative divide when it comes to designing a suitable methodology for their research, and select methods – quantitative, qualitative or a combination of the two – that best satisfy the needs of the specific research projects".

In the following section we describe the phases of research design included in our current project, which is adopting a mixed-method approach. The aim is to tease out the key research questions and to justify the rationale for such an approach.

19.5 THE RESEARCH AGENDA

The on-going research project seeks to understand current levels of GIS ▮
primary and secondary health care organisations, and to explore the hypothesi▮
are being under-utilised in the health sector. This ESRC-funded project has a
components including:

- Surveying levels and variations in GIS utilisation within the primary and
  health care sectors – building upon a previous cross sectional survey of
  Gould, 1992).

- Reviewing/survey types of GIS application being used in the light of rece▮
  in the nature of structure of the primary and secondary health care sectors.

- Gauging the importance of organisational factors in the widespread a▮
  such systems – such factors have been deemed to be important in other b▮
  medical informatics (*e.g.* Kaplan, 1997).

- Understanding reasons for variations in the use and wider implementati▮
  including the types of technical and organisational barriers that influe▮
  application.

- Investigating the implications of recent NHS policy changes on th▮
  geographically referenced data and GIS technology.

- Examining the nature and extent of intra-/inter-data exchanges within
  given the perceived advantages of GIS as an integrating technology an▮
  arching aim of encouraging inter-agency collaboration in '*Our Healthier* ▮

- Highlighting 'best-practice' case studies of health service organisations▮
  using GIS for addressing key policy tasks.

- Exploring the potential use of other NHS C&IT (communications and in▮
  technology) developments and initiatives for sharing health informa▮
  NHSNet, NHS direct and Internet-based GIS).

This research agenda seeks to capture and assess the impact of the recent ▮
central government to exploit the rich source of geographically referenced
within the NHS, to enhance *both* the delivery and monitoring of health care se▮
Lusignan *et al.*, 2000; Bell, 2000; Gibbs, 2000). This is manifest in the c
enabling organisations to translate and implement the goals of a new In▮
Strategy for the NHS, as outlined in The NHS Plan (Department of Health, ▮
more specifically, Information For Health (NHS Executive, 1998). In ad▮
would contend that the application of GIS in the health sector can, in turn, prov▮
bed for many of the issues surrounding 'joined-up government' debates which a▮
the use of GIS in other sectors such as local government and crime.

Interestingly, there is no mention of the use of GIS within recent policy ▮
despite well documented examples of the potential for GIS within the NHS
1991; Birkin *et al.*, 1996; Todd and Forbes, 1998; Gatrell and Loytonen, 1998▮
numerous academic projects that have shown the benefits of GIS for he▮
application tasks (Gatrell and Senior, 1999; Higgs and Gould, 2001). This c▮
examined in the light of more complex, integrated and pervasive informatio▮
coming on stream within the NHS and a relative proliferation in spatially ▮
disaggregate data sources (Bryant, 1998). Clearly, if NHS organisations ar▮
benefit from these holistic information systems and respond to recent g▮

objectives, new technologies such as GIS, must be fully harnessed to realise their potential.

Therefore, the current and potential use of GIS must be accurately measured within the emerging 'information-drive' context of the NHS. This is essential if Cummins and Rathwell's (1991) vision for the integration, manipulation and presentation of locational and spatial data within the NHS is to be realised. The new policy initiatives of *The NHS Plan* and *Information for Health*, as well as the promotion of 'joined-up' government, collaboration, and joint commissioning, have offered substantial encouragement for the enhanced use of GIS within the NHS (Higgs and Gould, 2001). This project will examine if such influences are having a significant impact on GIS usage in the health sector.

19.6 A MIXED-METHOD RESEARCH DESIGN

In a recent paper, Higgs and Gould (2001: 19) comment: "people wanting to survey current GIS usage within the NHS are faced with a difficult task as there is no single source of information". In the absence of such an accurate baseline of the current use of GIS within the NHS, the starting point for our research process involved the preparation of an extensive questionnaire-based survey, in order to provide a context for a wider exploration of the issues influencing the effective implementation of GIS.

In order to ensure that the questionnaire schedule was comprehensive, it was deemed essential to fulfil two key tasks prior to designing the questionnaire schedule and administering the survey. First, we reviewed the questionnaire schedules employed for previous studies of the use of GIS within the NHS (*e.g.* Gould, 1992), as well as questionnaire-based surveys of the use of GIS in other public sector contexts (*e.g.* Campbell and Masser, 1993; Church *et al.*, 1993; Gill *et al.*, 1999). This consolidated and informed our understanding of the breadth of issues under investigation as well as helping us to estimate the realistic levels of response we could expect from addressing relatively complex factors that may be influential in the effective implementation of GIS. Secondly, the questionnaire schedule was sensitised further following pilot discussions with GIS experts and with health-professionals, key policy makers and institutional actors with a broader knowledge of the issues involved (Table 19.1).

Table 19.1 Stages of Piloting and Consultations (Dates, Venues).

- An audience of academics with GIS and health-related interests at the GISRUK 2001 conference, University of Glamorgan, April 2001.

- A Department of Health Geographic Information Steering Group meeting, London, May 2001.

- A policy-oriented audience and GIS suppliers / users at an Association for Geographical Information Health Special Interest Group conference, University of Aston, Birmingham, June 2001.

- A pilot survey of Health Authorities in the Trent region, June 2001.

The questionnaire schedules were tailored for Health Authorities and co
in England, Wales, Scotland and Northern Ireland to take into account differ
directives and organisational structures. Other important suggestions inc
clarification and modification of questions (*e.g.* adding definitions), the incorp
additional questions (*e.g.* questions relating to GIS support or advice for Prir
Groups / Trusts), the availability of a web-based option for answering the que
and the degree of question overlap with previous surveys (*e.g.* Gould, 1991),
turn, permit some degree of time-series analysis (Smith *et al,* in preparation).

The second phase of the research process involved the deliver
questionnaire-based survey. This was posted out, and also made available for c
on the WWW, during June and July 2001 to appropriate Information Tec
Services personnel in all Health Authorities of England (95) and Wales (4
Boards in Scotland (15), and Health and Social Services Boards in Northern Ir
and all Health Trusts (469), and to the General Managers / Chief Executives fr
random sample of Primary Care Groups/Trusts across the UK (193), stratified
Authority Region. The sampling frame was constructed from Binley's (27
Spring 2001) databases for *'NHS Management'* and *'Primary Care Groups/T*
total 780 surveys were distributed.

The main purpose of the questionnaire was to provide standardised in
elicited from both open and closed-ended questions, which would provide ar
measure of current and potential uses of GIS, and would facilitate comparison
different respondents. It was essential that the questionnaire also flagged up fact
respondents perceived to hinder or enable use and uptake of GIS wi
organisation. Issues such as the importance of the commitment of key person
effectiveness of GIS implementation were sought at this stage – such factors ha
seen as important in other health contexts (Houghton, 2001). A further outco
survey was that 'best practice' examples of the use of GIS within the NHS
identified. This latter element was essential, in order to select appropriate cas
which could then be explored more fully via face to face qualitative interviews.

Based upon our pre-existing understanding of the factors which constrain
GIS use within the NHS, the questionnaire-based survey was divided into
sections. Section 1 sought to identify the diversity of institutional contexts v
NHS which may influence a range of current GIS uptake and uses. The aim ha
document the historical background of GIS uptake and uses; ide
presence/absence of a GIS strategy; establish whether the current GIS is fully or
and; tease out how the GIS is administered and maintained. The levels of i
inter- organisational usage of GIS have been examined in order to identify
organisations turn for GIS advice and analysis, and to examine levels of leade
political support for such initiatives. In Section 2 we attempted to reveal any fu
for GIS uptake and use, for example, by asking respondents if the organisatio
plans to modify their current GIS or purchase a new GIS. The aim of Section
pin down the tasks for which GIS is currently being used and to identify poli
uses of GIS. The projected use of GIS was also identified through subsidiary
as was the importance of recent official NHS documents/policy guidance
implications of GIS on the wider availability of health information. In Sec
specifically focused on the technical constraints that are often viewed as key c
hindering GIS within the NHS. Thus, we asked questions about the type of da
are being used for GIS purposes and the types of data that are being exchanged
organisations, and the factors that are limiting accessibility to datasets. Other

constraining factors were then considered in Section 5, such as the level of access to and types of training, the availability of advice, guidance and support for GIS implementation and use. Crucial here was the need to obtain the subjective and personal opinions of respondents regarding the types of organisation that they felt should provide training and guidance, which could then be followed up in face to face interviews in phase 2. The final section sought to corroborate earlier answers and illuminate the significance of factors which respondents had previously cited in their responses to questions posed about the types pf factors which enabled and constrained GIS uptake within their respective organisations. As Kaplan (1997: 97) suggests, with regard to wider studies examining the impact of technology in medicine, "evaluators often could benefit from multiple research methods". The approach we have taken therefore aims to build on a questionnaire approach to address these issues.

From a preliminary analysis of the survey results, 'best-practice' examples of GIS use within NHS organisations can be identified. These case studies will form the basis for the next phase of the research involving semi-structured interviews with key informants in a variety of organisational settings, building on findings of the initial survey. Qualitative data will be gathered and analysed (using grounded theory techniques) to permit a deeper understanding of some of the contingent social and cultural influences, and the dynamic processes underpinning the uptake and uses of GIS within different organisational contexts of the NHS. Examples of inter-related themes which will be investigated during the face to face interviews are outlined in Table 19.2.

Table 19.2 Examples of the themes under investigation in phase 2 of the project.

Individual issues	*Organisational issues*
• Knowledge and understanding	• Structural and organisation change
• Individual interests and motivations	• Corporate utility
• Early innovators/organisational 'champions'	• Organisational culture
• Previous experience and education	• Accountability
• Trained staff transferring to the private sector	
Policy issues	*Resource issues*
• Policy imperatives	• Staffing
• Local information/C&IT strategies	• System requirements
• Local and regional politics	• Educational/training needs
• Top-down and bottom-up pressure	• Communication
Data issues	
• Data commodification	
• Confidentiality	

19.7 DISCUSSION

It is envisaged that the mixed-method approach adopted for this research p
enhance previous studies of the use of GIS within the NHS, and provide unde
drawn from different levels of explanation. Whilst the extensive questionna
has contextualised and provided evidence of trends and regularities in GIS upta
across NHS organisations, it is anticipated that the intensive semi-structured
will provide a fuller understanding of the links between causal factors,
managerial commitment to GIS and organisational stability on GIS uptake an
use. As Kaplan (1997: 97) notes: "combining qualitative with quantitativ
allows for a focus on the complex web of technological, economic, organisa
behavioural issues", and adds to the robustness of research findings. Th
addressed in the second phase of the project.

Although we fully recognise that triangulating the different strands of th
process will be problematic, we would argue that it is highly beneficial. The
triangulation will provide complimentary empirical evidence, and both supp
contradict the findings from the questionnaire survey and interviews. Moreove
offer the opportunity to cross-check results and methods, and thus tease
meaningful inconsistencies which emerge, and reduce the risk of establishing
conclusions. This is likely to offer new insights into the levels of GIS usage i
NHS organisational contexts. Moreover, and in contrast to previous studies of
GIS within the NHS, we would assert that the mixed method approach will
rigorous exploration of the factors constraining GIS uptake and use wi
organisations, and is likely to facilitate greater confidence in the initial resea
will be important if policy makers are to acknowledge the findings from the res
take into account the factors which are influencing GIS implementation within t

In conclusion, this chapter has sought to draw attention to the benefits of
mixed methods approach to investigate the factors which enable or constrain tl
and use, of GIS within NHS organisations. The main benefits of this app
complimentarity, confirmation, reliability, increased confidence levels
illumination of meaningful contradictions and inconsistencies. This, in turn,
benefits for researchers examining the impacts of new technologies within ot
and private sector organisations.

19.8 ACKNOWLEDGEMENTS

The project is being funded by an ESRC research grant (R000223473). We ack
the support of the Department of Health GI Steering Group and the AGI H
Particular thanks to Ralph Smith, Paul White, Duncan Cooper, Ian Bullard and
Dunn for comments on the design of the questionnaire. However, the opinions
in this paper are those of the authors alone and may not necessarily be represe
the organisations listed above.

19.9 REFERENCES

Barlow, T., 2000, GIS and mapping in healthcare. *The British Journal of Healthcare Computing and Information Management,* **17**, pp. 42-44.

Bell, N., 2000, Realising the vision – making a difference. *The British Journal of Healthcare Computing and Information Management,* **17**, pp. 17-19.

Birkin, M., Clarke, G., Clarke, M. and Wilson, A., 1996, *Intelligent GIS: Location Decisions and Strategic Planning.* (New York: Wiley).

Brannen, J., 1992, *Mixing-methods: Quantitative and Qualitative Research.* (Aldershot: Avebury).

Bryant, J., 1998, The importance of human and organisational factors in the implementation of computerised information systems: an emerging theme in European healthcare. *The British Journal of Healthcare Computing and Information Management,* **15**, pp. 27-30.

Burns, S., 1996, Mapping Healthcare: What's the Point? *The British Journal of Healthcare Computing and Information Management,* **13**, pp. 37-39.

Cooper, D., 2000, Developing GIS within the NHS – obstacles and possible solutions. *The British Journal of Healthcare Computing and Information Management,* **17**, pp. 31-33.

Campbell, H. and Masser, I., 1993, The impact of GIS on local government in Great Britain. In *Geographical Information Handling – Research and Applications,* edited by Mather, P.M. (London: Wiley), pp. 273-286.

Church, A., John, S., Shepherd, W., Frost, D. and Macmillan, A., 1993, An information system for the Eastern Thames Corridor. In *Geographical Information Handling – Research and Applications,* edited by Mather, P.M., (London: Wiley), pp. 287-324.

Cummins, P. and Rathwell, T., 1991, *Geographical Information Systems and the National Health Service Education, Training and Research Public*ation. (London: Association of Geographic Information).

Department of Health., 2000, *The NHS Plan. A Plan for Investment, a Plan for Reform.* (London, HMSO, White Paper).

Findlay, A.M. and Li, F.L.N., 1999, Methodological issues in researching migration. *Professional Geographer,* **51**, pp. 50-59

Gatrell, A.C., 2001, *Geographies of Health: An Introduction.* (Oxford: Blackwell).

Gatrell, A.C. and Loytonen, M., 1998, *GIS and Health.* (London: Taylor and Francis).

Gatrell, A.C. and Senior, M., 1999, Health and health care applications. In *Geographical Information Systems: Principles, Techniques, Applications and* Management, edited by Longley, P., Maguire, D., Goodchild, M. and Rhind, D., (London: Wiley), pp. 925-938.

Gill, S., Higgs, G. and Nevitt, P., 1999, GIS in planning departments: preliminary results from a survey of local authorities. *Planning and Policy in Practice,* **14**, pp. 341-361.

Gould, M., 1992, The use of GIS and CAC by health authorities: results from a postal questionnaire. *Area,* **24**, pp. 391-401.

Graham, E., 1999, Breaking out: the opportunities and challenges of multi-method research in population geography. *Professional Geographer,* **51**, pp. 76-89.

Higgs, G. and Gould, M., 2000, Commentary: healthcare commissioning, the modern NHS, and geographical information systems. *Environment and Planning A,* **32**, pp. 1905-1908.

Higgs, G. and Gould, M., 2001, Is there a role for GIS in the 'New NHS'. *Health and Place,* **72**, pp. 247-259.

Houghton, F., 2001, Health GIS in Ireland – lessons from New Zeal.
Geography, **34**, pp. 96-97.

Kaplan, B., 1997, Addressing organisational issues into the evaluation o
systems. *Journal of the American Medical Informatics Association*, **4**, pp. 9.

Lusignan, S., Mimnagh, C., Kennedy, J. and Peel, V., 2000, Alignment of infor
health with the NHS plan - a case for substantial investment and reform. *T*
Journal of Healthcare Computing and Information Management, **17**, pp. 28.

Martin, D., 1996, *Geographic Information Systems: Socioeconomic Ap*
(London: Routledge).

McKendrick, J., 1999, Multi-method research: an introduction to its appl
population geography. *Professional Geographer*, **51**, pp. 40-50.

NHS Executive., 1998, *Information for Health: An Information Strategy for th*
NHS, (London: NHS Executive).

Philip, L.J., 1998, Combining quantitative and qualitative approaches to social r
human geography – an impossible mixture? *Environment and Planning .*
261-276.

Reeve, D. and Petch, J., 1999, *GIS Organisations and People: A Socio-*
Approach. (London: Taylor and Francis).

Smith, R., 1999, Co-ordinating healthcare – taking a geographic perspective. *T*
Journal of Healthcare Computing and Information Management, **16**, pp. 40.

Smith, R. and Jarvis, C., 1998, Just the Medicine. *Mapping Awareness*, **12**, pp. .

Smith, D.P., Higgs, G. and Gould, M., 2001, Clarifying the current and poten
GIS within NHS organisations: optimism or pessimism? *The British J*
Healthcare Computing and Information Management, **18**, pp. 24-26.

Smith, D.P., Gould, M. and Higgs, G., in prep, *Surveying the uses of GIS in the*
ten years on.

Sporton, D., 1999, Mixing methods in fertility research. *Professional Geogre*
pp. 68-76.

Taylor, G. and Allgar, G., 1997, The creation of an enhanced dataset from
medical records using a relational management system. *The British J*
Healthcare Computing and Information Management, **14**, pp. 32-34.

Todd, P. and Forbes, H., 1998, Seeing statistics more meaningfully: a spatia
package for epidemiology. *The British Journal of Healthcare Comp*
Information Management, **15**, pp. 32-34.

Wrigley, N., 1991, Market-based systems of health-care provision, the NHS
Geographical Information Systems. *Environment and Planning A*, **23**, pp. 5-

Index

Printed and bound by CPI Group (UK) Ltd, Croydon, CR0 4YY

24/10/2024

01778285-0004